▶ Applying Respondent Driven Sampling to Migrant Populations

Other Palgrave Pivot titles

G. Douglas Atkins: **T.S. Eliot and the Fulfillment of Christian Poetics**

Guri Tyldum and Lisa G. Johnston (editor): **Applying Respondent Driven Sampling to Migrant Populations: Lessons from the Field**

Shoon Murray: **The Terror Authorization: The History and Politics of the 2001 AUMF**

Irene Zempi and Neil Chakraborti: **Islamophobia, Victimisation and the Veil**

Duggan, Marian and Vicky Heap: **Administrating Victimization: The Politics of Anti-Social Behaviour and Hate Crime Policy**

Pamela J. Stewart and Andrew J. Strathern: **Working in the Field: Anthropological Experiences across the World**

Audrey Foster Gwendolyn: **Hoarders, Doomsday Preppers, and the Culture of Apocalypse**

Sue Ellen Henry: **Children's Bodies in Schools: Corporeal Performances of Social Class**

Max J. Skidmore: **Maligned Presidents: The Late 19th Century**

Lynée Lewis Gaillet and Letizia Guglielmo: **Scholarly Publication in a Changing Academic Landscape**

Owen Anderson: **Reason and Faith in Early Princeton: Piety and the Knowledge of God**

Mark L. Robinson: **Marketing Big Oil: Brand Lessons from the World's Largest Companies**

Nicholas Robinette: **Realism, Form and the Postcolonial Novel**

Andreosso-O'Callaghan, Bernadette, Jacques Jaussaud, and Maria Bruna Zolin (editors): **Economic Integration in Asia: Towards the Delineation of a Sustainable Path**

Umut Özkırımlı: **The Making of a Protest Movement in Turkey: #occupygezi**

Ilan Bijaoui: **The Economic Reconciliation Process: Middle Eastern Populations in Conflict**

Leandro Rodriguez Medina: **The Circulation of European Knowledge: Niklas Luhmann in the Hispanic Americas**

Terje Rasmussen: **Personal Media and Everyday Life: A Networked Lifeworld**

Nikolay Anguelov: **Policy and Political Theory in Trade Practices: Multinational Corporations and Global Governments**

Sirpa Salenius: **Rose Elizabeth Cleveland: First Lady and Literary Scholar**

Sten Vikner and Eva Engels: **Scandinavian Object Shift and Optimality Theory**

Chris Rumford: **Cosmopolitan Borders**

Majid Yar: **The Cultural Imaginary of the Internet: Virtual Utopias and Dystopias**

Vanita Sundaram: **Preventing Youth Violence: Rethinking the Role of Gender and Schools**

Giampaolo Viglia: **Pricing, Online Marketing Behavior, and Analytics**

Nicos Christodoulakis: **Germany's War Debt to Greece: A Burden Unsettled**

palgrave▶pivot

Applying Respondent Driven Sampling to Migrant Populations: Lessons from the Field

Edited by

Guri Tyldum
Fafo Institute for Applied International Studies, Norway

and

Lisa G. Johnston
Tulane University and University of California, USA

DOI: 10.1057/9781137363619.0001

Selection and editorial content © Guri Tyldum and Lisa G. Johnston 2014
Individual chapters © the contributors 2014

All rights reserved. No reproduction, copy or transmission of this publication may be made without written permission.

No portion of this publication may be reproduced, copied or transmitted save with written permission or in accordance with the provisions of the Copyright, Designs and Patents Act 1988, or under the terms of any licence permitting limited copying issued by the Copyright Licensing Agency, Saffron House, 6–10 Kirby Street, London EC1N 8TS.

Any person who does any unauthorized act in relation to this publication may be liable to criminal prosecution and civil claims for damages.

The authors have asserted their rights to be identified as the authors of this work in accordance with the Copyright, Designs and Patents Act 1988.

First published 2014 by
PALGRAVE MACMILLAN

Palgrave Macmillan in the UK is an imprint of Macmillan Publishers Limited, registered in England, company number 785998, of Houndmills, Basingstoke, Hampshire RG21 6XS.

Palgrave Macmillan in the US is a division of St Martin's Press LLC, 175 Fifth Avenue, New York, NY 10010.

Palgrave Macmillan is the global academic imprint of the above companies and has companies and representatives throughout the world.

Palgrave® and Macmillan® are registered trademarks in the United States, the United Kingdom, Europe and other countries.

ISBN: 978–1–137–36362–6 EPUB
ISBN: 978–1–137–36361–9 PDF
ISBN: 978–1–137–36360–2 Hardback

A catalogue record for this book is available from the British Library.

A catalog record for this book is available from the Library of Congress.

www.palgrave.com/pivot

DOI: 10.1057/9781137363619

Contents

List of Figures	ix
List of Tables	x
Acknowledgments	xi
Definitions of RDS Terminology	xii
Notes on Contributors	xiv

	Introduction *Lisa G. Johnston and Guri Tyldum*	1
	A need for data about migration	2
	RDS and migrant populations	3
	Aims and structure of this book	5
1	Sampling Migrants: How Respondent Driven Sampling Works *Lisa G. Johnston*	9
	Introduction	10
	How RDS works	11
	RDS assumptions	13
	Conclusion	16
2	RDS and the Structure of Migrant Populations *Jon Horgen Friberg and Cindy Horst*	17
	Introduction	18
	Why RDS is well suited to studying migrant populations	18

	Target populations and naturally occurring social groups – common problems	21
	Bottlenecks and clustering	23
	Getting to know the study population	24
	Conclusion	26
3	Measuring Personal Network Size in RDS *Lisa G. Johnston, Leila Rodriguez and Joanna Napierala*	27
	Introduction	28
	The PNS variable and why we need it	28
	Constructing the personal network size question(s)	29
	Clear definition of the target population	29
	The meaning of "knowing" someone	30
	Geographic boundary	30
	Time frame in which the respondent has seen their peers	31
	Measuring PNS	33
	Eliciting PNS by sub-group	33
	Training staff	33
	PNS of zero, outliers and coarsened data	34
	Temporal impacts	35
	Conclusion	36
4	Initiation of the RDS Recruitment Process: Seed Selection and Role *Agnieszka Kubal, Inna Shvab and Anna Wojtynska*	37
	Introduction	38
	Strategic selection of seeds	38
	Identifying seeds	41
	Number of seeds	44
	How seeds work – script for recruitment	46
	Conclusion	47
5	Deciding on and Distributing Incentives in RDS *Guri Tyldum, Leila Rodriguez, Ingunn Bjørkhaug and Anna Wojtynska*	49
	Introduction	50
	Motivating survey respondents to take part	50
	Primary incentive	51
	Secondary incentive	51

Determining the type and value of the incentive	52
Compensating for time use in line with average salaries for the group	52
Stratified incentives	54
The impact of incentives that are too high or too low	55
Non-monetary incentives	56
RDS without material incentives	56
Making participation a positive experience	57
Organizing the distribution of incentives	57
The ethics of incentives	59
Conclusion	61

6 **Formative Assessment, Data Collection and Parallel Monitoring for RDS Fieldwork**   62
Jane Montealegre, Antje Röder and Rojan Ezzati

Introduction	63
Planning and formative assessment	66
Survey sites	67
Staffing	69
Survey coupons	70
Data collection start date	74
Data collection and parallel monitoring	74
Initiating data collection	74
Methods for parallel monitoring	75
Addressing slow recruitment	77
Addressing rapid recruitment	78
Masquerading and repeat respondents	79
Ending RDS	80
Ethical considerations	81
Conclusion	82

7 **Analyzing Data in RDS**   84
Lisa G. Johnston and Renee Luthra

Introduction	85
A need for special analysis of RDS data	85
Which software to use when analyzing RDS data	86
Deciding which estimator to use	87
Variance in RDS analysis	90
Assessing bias in RDS analysis	90

Seed dependence	90
Homophily	92
Differential recruitment activity	94
Analyzing bottlenecks	94
Exporting weights for multivariate analysis	96
Reporting RDS findings	97
Using RDS findings to impact policy	98
Conclusion	99
Appendix I: Summary of RDS Surveys Referenced	101
Central American Women in Houston	101
Foreign migrants in Ukraine	102
Migrants in Warsaw, 2010 and 2012	102
Nigerians in New York City	104
Polonia in Oslo, 2006 and 2010	104
Polonia in Reykjavik	105
Polonia in Dublin	106
Sub-Saharan Africans in Morocco	107
THEMIS	108
SCIP project studies	109
Appendix II: Overview of the Development of the Different RDS Estimators, Their Specific Features and the Software Available for Their Use	110
References	114
Index	124

List of Figures

1.1	Recruitment chain in RDS. Illustration of seed and waves	12
1.2	Equilibrium for males and females	15
3.1	Personal network size question	32
3.2	Histogram of personal network size by wave	35
4.1	Initial seed bias in recruitment of married and unmarried respondents	39
4.2	Recruitment of men and women with different outcomes of samples achieved from one seed	45
6.1	Overview of respondent flow. Example	71
6.2	Survey coupon (front and back)	72
6.3	Survey coupon with section to be kept by recruiter (front and back)	73
6.4	Pattern of recruitment, by week	75
6.5	Paper and pencil diagram to monitor cross-site recruitment	76
6.6	Diagram of recruitment, by country of origin	76
6.7	Pattern of rapid recruitment, by week	79
7.1	Equilibrium points for two variables	91
7.2	Homophily for females in sample and population	93
7.3	Bottlenecks within neighborhoods	95

List of Tables

7.1	Sample and population estimates of gender distribution	86
7.2	Example of output from RDS-Analyst	89
7.3	Description of recruitment by seeds	97

Acknowledgments

This volume was made possible with economic support from the Norwegian Directorate of Immigration, the Norwegian Research Council, the University of Cincinnati and Fafo Institute for Applied International Studies. We would like to extend our sincere gratitude to all of the contributors to this volume for their involvement and contributions. A particular thank you goes to Leila Rodriguez, who provided additional assistance in finalizing this book, and to Renee Luthra, who came up with the idea of this volume.

Definitions of RDS Terminology

Bottleneck – The absence of personal links between different sub-groups within the target population.

Clustering – Where parts of a population are much more densely connected than others.

Coupon – An invitation that a respondent can give to other individuals, ideally members of the target population, to take part in a survey. Coupons have a unique number, linking them with the recruiter.

Equilibrium – The point in recruitment when the proportion of a sample characteristic is assumed to be independent from the characteristics of the seeds. This is based on an assumption that there is a point (by wave or respondents) in the recruitment chain whereby the proportions of each variable no longer change despite the chain accumulating more waves or respondents.

Homophily – The tendency for individuals to purposefully recruit others with similar characteristics to themselves, rather than recruiting randomly from the network of characteristics. The term is also used to describe similar characteristics existing in relationships of target populations.

Recruit – An individual who receives a coupon from a survey respondent, and who agrees to enroll in the survey.

Recruiter – An individual who recruits someone by giving them a coupon, expecting them enroll in the survey.

Recruitment chain – The set of all respondents linked to a specific seed. The seed and the waves make up a recruitment chain (also sometimes called a 'recruitment tree').

Recruitment matrix – A matrix (table of columns and rows) of the characteristics of both the recruiter and the recruited. For instance, the matrix for the rows will present the recruiters' characteristics (i.e., males and females), and the columns will present the characteristics of the recruited (i.e., males and females). The recruitment matrix is used in several of the RDS estimators.

Personal network size (PNS) – The number of reciprocal relationships a respondent has to other members of the target population.

Primary incentive – The money, goods, and/or services provided to respondents for completing the main interview.

Secondary incentive – The money, goods, and/or services provided to respondents for each new respondent they are able to recruit.

Seed – A member of the target population who is recruited by a researcher to start a recruitment chain. All RDS studies begin with the selection of least one seed.

Wave – Indicates the number of links between a respondent and individuals recruited after the seed (wave 0). When there is more than one wave, there is a recruitment chain.

Notes on Contributors

Ingunn Bjørkhaug specializes in development studies and is a doctoral researcher at Fafo Institute for Applied International Studies in Norway. She has conducted a number of studies in post-conflict settings on the subject of displacement, gender-based violence and ex-combatants in Colombia and Liberia. She has been co-responsible for the implementation of RDS surveys in Sierra Leone and Liberia.

Rojan Ezzati is a sociologist and doctoral researcher based at the Peace Research Institute Oslo (PRIO). Her research interests include how societies respond to terror, development processes of the nation, as well as international migration and transnational activities among migrants. As part of the THEMIS team, she was co-responsible for the implementation of the RDS surveys in the Norwegian component of the project.

Jon Horgen Friberg is a sociologist and Research Fellow at the Fafo Institute for Labour and Social Research in Norway. His main areas of interest are migration, labor markets and ethnic relations. He has been co-responsible for the implementation and analysis of two RDS surveys among Poles in Oslo, and has published extensively on the material from these surveys.

Cindy Horst is Research Professor in Migration and Refugee Studies at the Peace Research Institute Oslo (PRIO). Her current research interests include mobility in conflict, diaspora, humanitarianism, refugee protection, transnational civic engagement, and theorizing on social transformation. She is particularly interested in methodological innovations that allow for critical and ethically conscious research engagement, through shared anthropology and multi-sited ethnography.

Lisa G. Johnston is an epidemiologist and independent consultant with affiliations at the University of California, San Francisco, Global Health Sciences and Tulane University School of International Public Health and Tropical Medicine. Dr. Johnston provides training and technical assistance worldwide on how to prepare, implement, monitor and analyze data from RDS studies and has conducted hundreds of RDS surveys on migrants and other hard-to-reach populations. She has authored/co-authored roughly 50 journal articles and reports on surveys using adaptive sampling methods, and has written training manuals for planning, implementing and analyzing data from RDS surveys. For more information see her website at: www.lisagjohnston.com.

Agnieszka Kubal is a research fellow at the Centre for Socio-Legal Studies, University of Oxford, and a research fellow at Wolfson College, University of Oxford. Her research interests encompass migrants' legal incorporation, the rights-citizenship nexus, questions of legality and semi-legality, social theory and comparative legal culture. As part of the THEMIS team she was responsible for the implementation of RDS surveys with Brazilians, Moroccans and Ukrainians in London in 2012.

Renee Luthra is a sociologist and research fellow at the Institute for Social and Economic Research at the University of Essex. Her work, funded by the Russell Sage Foundation, the Spencer Foundation, and Norface, examines the integration of the second generation and intergenerational mobility in immigrant families, as well as the socio-cultural outcomes of recently arrived immigrants in Europe.

Jane Montealegre is an instructor at the Dan L. Duncan Cancer Center at Baylor College of Medicine in Houston, Texas. She is a behavioral epidemiologist with an interest in sexual and healthcare utilization behaviors among migrant populations in the context of HIV and human papilloma virus infections. Her dissertation research on the effectiveness of RDS among undocumented Central American immigrant women in the U.S. was the first to evaluate the use of RDS to recruit undocumented immigrants for HIV behavioral research.

Joanna Napierala is an economist and research fellow at the Centre of Migration Research, University of Warsaw. She was involved in the application of RDS sampling in studies of Polish migrants in Oslo, Copenhagen and Reykjavik and of Ukrainians, Belarusians and Russians in Warsaw. She co-authored the publication on evaluating the effectiveness of the RDS sampling method in the study on Ukrainians.

Antje Röder is Ussher Assistant Professor in Migration at the Department of Sociology, Trinity College Dublin. Her main research interest is the intersection of host, origin and individual characteristics in migrants' social, cultural and economic integration in European societies. She has been involved in the Polonia in Dublin project, which was the first RDS study of Polish migrants in Ireland, and teaches research methodology at both undergraduate and postgraduate levels.

Leila Rodriguez is Assistant Professor in the Department of Anthropology at the University of Cincinnati. She is also a research affiliate of the Central American Population Center at the University of Costa Rica. With a dual PhD in Anthropology and Demography, her research utilizes a mixed methods approach to the study of socioeconomic aspects of international migration.

Inna Schvab is senior research officer and fieldworker at ICF 'International HIV/AIDS Alliance in Ukraine'. Since 2007 she has coordinated numerous national integrated bio-behavioral RDS studies among hidden populations vulnerable to HIV in Ukraine (up to 30 cites), including Ukrainian labor migrants with experience of working abroad (2010), foreign migrants in Ukraine (2013), people who inject drugs (2007, 2008, 2009, 2011, 2013), female sex workers (2008, 2009, 2011), and men who have sex with men (2007, 2009, 2011, 2011).

Guri Tyldum is a sociologist and research fellow at Fafo Institute for Labour and Social Research in Norway. She has broad experience in sampling and survey implementation, with hands-on experience from both large-scale household surveys in countries of transition and development, as well as smaller surveys on rare and elusive populations. Her current research interests include the dynamics of labor migration, migration theory and human trafficking, as well as ethical and methodological challenges in studies of marginalized populations.

Anna Wojtynska is a doctoral student at the Faculty of Social and Human Sciences at the University of Iceland. Her research project is about the transnational practices of Polish migrants in Iceland. In 2010, she coordinated the RDS survey in Reykjavik, carried out by CIRRA (Center for Immigration Research at the Reykjavik Academy), part of the 'Mobility and Migrations at the Time of Transformation: Methodological Challenges' project.

palgrave▶pivot

www.palgrave.com/pivot

Introduction

Lisa G. Johnston and Guri Tyldum

Abstract: This introduction illustrates the need for increased migration research, as more and more people cross national and international borders, and labor markets become increasingly reliant on migrant labor. Aside from the nature of many migrant populations being hard to access and measure due to issues of inclusion, identification, motivation and trust, there are few sampling methods available to provide representative information about migrants' living conditions, patterns of movement, discrimination, health, access to social services and other important data. Respondent driven sampling (RDS), a modified chain referral method, has been extremely successful in acquiring reliable data on hard-to-reach populations, mostly for public health research. However, over the past few years, RDS has been used effectively to sample migrants. In this chapter, we provide a brief overview of how RDS can address many of the challenges in sampling migrant populations and provide a list of the existing surveys that have used RDS to sample migrant populations. We further outline how the idea and implementation of this book came about and provide a brief introduction to each of the chapters contained within.

Tyldum, Guri and Lisa G. Johnston, eds. *Applying Respondent Driven Sampling to Migrant Populations: Lessons from the Field.* Basingstoke: Palgrave Macmillan, 2014.
DOI: 10.1057/9781137363619.0007.

A need for data about migration

We live in an age of migration (Castles & Miller, 2009). About 3.2% of the world's population do not reside in the country of their citizenship (UNDESA, 2013), and despite steadily increasing restrictions on migration worldwide, people continue to go abroad for a variety of reasons. The development of new transport and communication technologies, together with rapidly expanding transnational bonds, which tie people together across national boundaries, have made migration an option that is increasingly considered in various phases of life: to escape poverty or violence, find work, study, join family members who have already migrated, find love, see the world, or start a new life (King, 2002). For governments in immigrant receiving countries, the regulation of immigration flows and the integration of newcomers have become a central part of the political agenda, as migration is increasingly made relevant for policies on poverty alleviation, labor market regulations, the provision of health and educational services, and national security (UNHCR, 2013). In many countries of emigration, remittances have become an important part of gross domestic products, and migrant communities abroad play an important role in the social and political development of both sending and receiving countries worldwide. Industrial societies across the globe are becoming increasingly multicultural, and many economies are growing dependent on migration as sectors of the labor market, such as agriculture, construction, cleaning and care for children and the elderly, come to rely on migrant labor. The presence of migrants has sparked fierce debates over issues related to national identity, welfare state sustainability, and social cohesion.

Debates on migration-related policies are often confrontational and polarized, and tend to be shaped by the perceptions, ideas and concerns produced by media coverage, rather than academic knowledge (McKenzie & Marcin, 2007). The public image of migration is one of intense drama. We are confronted with images of boat refugees drowning off the coast of Lampedusa, the militarized US-Mexican border, Southeast Asian women freed from trafficking, and angry disfranchised youth from North Africa burning cars in the suburbs of European cities. The typical everyday lives of migrants are less likely to be reported in the news. Migrants are often marginalized in their countries of residence, and are often absent from public discourse. Much is still unknown about the processes and motivations that lie behind their mobility, as well as the mechanisms that lead to marginalization or success in their new location. And as practices and flows of migration are continually changing, what we knew about migration yesterday, might not still be the case today (King, 2002). This creates a continuous demand for accurate data and sound research in the migration field, to develop well-targeted policies, monitor change and to encourage a knowledge-based public debate on migration and its consequences.

RDS and migrant populations

Migrants make up a diverse category of persons who share the common factor of having moved within or between countries at some point. Within this group are migrants who are long-term residents in a receiving country with or without citizenship, refugees who may or may not wish to go back if the opportunity arose, people in irregular administrative situations (undocumented or "illegal" immigrants), short-term labor migrants and students. Another variation is that people can migrate alone, in groups, or with family. But in spite of their diversity, migrants share certain characteristics that make them particularly difficult to survey, because migrant populations introduces problems of *inclusion, identification, access, motivation* and *trust in the data production process* particular to this group.

The challenges of *inclusion* concerns mainly the way migrants tend to be only partially included in official statistics in countries of origin and destination (see Chapter 1 for more detailed discussion on this). Specialized surveys targeting migrants are also challenging, as there are few sampling frames available that enable us to *identify* migrants (or particular sub-groups of migrants) and distinguish them from non-migrants. The challenges of *access* are produced by, for instance, linguistic barriers, or problems finding migrants at home, if they work long hours. Limited spare time due to long working hours will also render them less *motivated* to spend time participating in a survey. Challenges of marginalization, racism and xenophobia, can make minority groups distrustful of persons or institutions representing the majority. Others will have good reason to wish to control the information collected about them and will have fears of revealing their identity. Thus, *lack of trust* may prevent some migrant groups from providing information to an interviewer.

Respondent Driven Sampling (RDS) addresses many of the challenges associated with sampling migrants. It does not rely on a pre-existing sampling frame, but uses the respondent's social networks to identify, recruit, and build trust among potential participants. It draws on the systematic use of incentives and peer-pressure to motivate participation and recruitment. In addition, RDS field organization can easily be adapted to improve access to particular linguistic groups or to establish interview times convenient for respondents. Finally, the coupon-based recruitment system enables fully anonymous participation (see Chapter 1 for more details).

It should, however, be stressed that in precision and variance, RDS can hardly compete with classical probability sampling designs. If relatively good sampling frames are available or can be constructed, and there are no major problems of trust, access or identification, traditional survey sampling is optimal. However, if convenience samples are the only alternative, then RDS is a better option. Therefore, as a methodology, RDS has increasingly gained popularity over the last few years in studies of migrant populations (Johnston & Malekinejad, 2014).

In Europe, RDS became an important methodology for accessing new Polish immigrants in Western European capital cities, such as Oslo (Friberg & Tyldum, 2007; Friberg & Eldring, 2011), Copenhagen (Hansen and Hansen, 2009), Dublin (Mühlau et al., 2011), and Reykjavík (Þórarinsdóttir & Wojtynska, 2011). In 2012–2013 a comparative survey of immigrants from Brazil, Morocco and Ukraine, living in the Netherlands, Norway, Portugal and the UK, was conducted (THEMIS, 2011). In 2011 attitudes about female circumcision among Somali immigrants in Oslo were studied with the help of RDS (Gele et al., 2012), while the Socio-Cultural Integration Processes among New Immigrants in Europe (SCIP) project used RDS to study integration trajectories among Poles and Pakistanis in London (Luthra et al., 2013). Examples using RDS to collect data from migrants in the United States include a 2004 survey of Latino first and second generation immigrant men in Chicago (Ramirez-Valles et al., 2005), a 2007 survey of Nigerian immigrant entrepreneurs in New York City (Rodriguez, 2009), a 2008 survey of unregulated workers, some of whom were migrants, in Chicago, Los Angeles and New York City (Milkman et al., 2010; Bernhardt et al., 2009), a 2010 survey of Central American Women in Houston (Montealegre et al., 2011; Montealegre et al., 2012), and a 2009 survey of undocumented workers in San Diego (Zhang, 2012). In the border zone between Thailand and Cambodia RDS was used to study an outbreak of malaria among migrant workers and (Khamsiriwatchara et al., 2011), and most recently, RDS was used to understand the HIV prevalence and risk factors among Anglophone and Francophone sub-Saharan African migrants living in Rabat, Morocco (Johnston, 2013a) and female migrants living in Cape Town, South Africa (Townsend et al., 2014).

RDS has previously been most widely used for epidemiological surveys to monitor HIV prevalence and risk behaviors in populations at higher risk of HIV, such as sex workers, men who have sex with men and people who inject drugs (Johnston et al., 2008; Malekinejad et al., 2008; Montealegre et al., 2013; Johnston, 2013b). Much information can be adapted and utilized directly from the wide range of manuals, materials and published literature based on these surveys. However, for various reasons, RDS surveys in migrant populations are not entirely comparable to HIV high-risk populations. The focus of this book is to explore the unique characteristics of migrant populations, to critically review challenges encountered by researchers when using RDS among migrants, and to provide lessons learned from researchers' experiences using RDS in migrant populations.

Aims and structure of this book

This book is the collaborative effort of a group of 15 researchers from various disciplines, including sociology, economics, public health, epidemiology and anthropology. Each of the authors has experience in using RDS in a migrant population. The idea for this book originated out of a workshop for researchers using RDS to survey migrant populations held at the University of London in 2011. We realized that little had been written on the practical challenges of data collection and analysis for RDS for this group, and the book is written in response to this, summarizing our experiences with RDS and discussing some of the key challenges encountered and the solutions introduced to address migrants. Each chapter is meant to stand alone so that readers can choose to read them in the order they are presented or to jump to chapters that are of most interest to them.

The book is collaborative in the sense that the researchers involved have reviewed and provided feedback to each other's chapters, which are also enhanced by the researchers sharing their own examples from fieldwork. This collaboration has allowed each chapter to draw on a variety of experiences from RDS surveys around the world. A number of examples and illustrations in the book were previously unpublished in their present form. The RDS surveys they build on are described in more detail in Appendix I. For those interested in reading more about specific surveys, available publications are listed in the same appendix.

The chapters following this Introduction will describe the RDS methodology, using examples from surveys of migrants. Chapter 1, *Sampling Migrants: How Respondent Driven Sampling Works*, introduces RDS methodology and the assumptions on which it is based. The subsequent chapters are presented in the order that a study would probably be planned and implemented. For instance, most RDS surveys require knowledge of the target population, their network properties and their acceptance of the RDS method, prior to initiating data collection. Chapter 2, *RDS and the Structure of Migrant Populations*, discusses the basic characteristics of migrant populations, and how information about a population can be used to determine whether RDS is an appropriate sampling method.

An essential part of RDS methodology is to measure each respondent's social network size in order to determine the probabilities of inclusion. Understanding the importance and use of this question is vital to ensure that respondents respond with enough accuracy to weight the data. Chapter 3, *Measuring Personal Network Size in RDS*, discusses the difficulties in estimating respondents' personal network sizes due to the fluidity and seasonal variation in migrants' networks, and gives practical advice on how to construct the personal network size question for migrant populations.

The careful selection of *seeds*, the non-randomly selected participants who initiate recruitment, will increase the probability of success of an RDS survey. Having seeds with diverse characteristics and who know a variety of people from the population being sampled will help to ensure that all sub-groups are included in the final sample. It will also speed up the attainment of equilibrium. Having too many seeds may result in short recruitment chains by the time the sample size is reached. Having too few seeds may result in slow recruitment and, if some seeds do not recruit anyone, may result in having to add more seeds after recruitment begins. In Chapter 4, *Initiation of the RDS Recruitment Process: Seed Selection and Role*, readers learn about monitoring and training seeds, and are provided with examples of both successful and unsuccessful seed selections.

One of the most difficult decisions to make for an RDS survey is determining the type and level of incentive that will be attractive enough for all sub-groups in a population to participate. If the incentive is too low, the sample could be biased by excluding key sub-groups and if the incentive is too high, the sample could be biased by encouraging people to sell or barter coupons, or to pretend to be a member of the population when they are not. Incentives can be in the form or money, gifts or other items of value. In Chapter 5, *Deciding on and Distributing Incentives in RDS*, readers learn how to identify and respond to these issues, they receive information and examples about monitoring incentives, making decisions about the appropriate incentive size and type, and the ethical issues to consider when using incentives.

Fieldwork organization in an RDS survey requires both thorough planning and the flexibility to adapt the design as you go along. These requirements demand that the questionnaire used for data collection is supplemented with formative assessment and parallel monitoring, allowing you to create an RDS design that is optimal for your population group, and to identify and address problems if they occur after data collection starts. Chapter 6, *Formative Assessment, Data Collection and Parallel Monitoring for RDS Fieldwork*, exposes readers to formative assessment strategies and different field set-up scenarios, including logistical decisions related to staffing and survey site selection, hours of operation, field monitoring and qualitative assessment, and considerations on when and how to end RDS.

Before initiating an RDS survey, researchers need to understand the data analysis aspect of RDS, as many of the decisions made during planning and data collection may impact analysis. Learning that RDS data requires special analysis may seem daunting to most researchers, especially those with little statistical experience, and doing analysis on a network and presenting findings that represent a network can be confusing at first. In addition, there have been a number of recent advancements in the development of the estimators used to analyze RDS data, as our understanding of how social science statistics and RDS assumptions merge with real-world situations of using RDS in networked populations. Chapter 7, *Analyzing Data in RDS*, provides readers with a description of the RDS estimators and the software in which these estimators are found. In addition, this chapter introduces some key concepts for RDS analysis, and types of biases affecting RDS samples, as well as some of the challenges researchers face during the analysis of their RDS data of migrant populations.

There are two appendixes to the book. Appendix I presents the surveys on which the experiences described in this book are based, while Appendix II presents an overview of the various estimators and software available for RDS analysis.

1
Sampling Migrants: How Respondent Driven Sampling Works

Lisa G. Johnston

Abstract: *Migrants are usually considered hard-to-reach for research purposes. This chapter provides an overview of RDS and describes how and why it is superior to other common types of sampling methods used to sample hard-to-reach populations. RDS uses the social network properties of populations to enable peers to recruit their peers. Although the method may appear to be straightforward, there are several important assumptions upon which recruitment and analysis are based. This chapter lays the foundation upon which the other chapters are based.*

Tyldum, Guri and Lisa G. Johnston, eds. *Applying Respondent Driven Sampling to Migrant Populations: Lessons from the Field*. Basingstoke: Palgrave Macmillan, 2014. DOI: 10.1057/9781137363619.0008.

Introduction

Migrants comprise populations that are considered hard-to-reach, making them difficult to sample using traditional probability methods. First, these populations lack the sampling frames needed to accurately determine the probability that each person has a chance of being selected for a sample (Kalton, 2001; Kalton & Anderson, 1986). These populations are also hard-to-reach because of language differences, time constraints due to long or irregular working hours, lack of trust due to marginalization, racism and stigma (IOM, 2001), and due to illegality as a result of being employed in shadow economies and having an irregular administrative status.

Data from demographic measurement systems, such as population censuses and registers, border admission, duration, work permit or other administrative systems are inconsistent across countries (including sending and receiving countries), and capture limited information on less hidden portions of migrant populations (Groenewold & Bilsborrow, 2008). Efforts to capture information from sizable samples of migrant populations have relied on household surveys, and on targeted and snowball sampling methods (McKenzie & Mistiaen, 2009). Household surveys using traditional random sampling or random digit dialing are capable of gathering representative data but are expensive, time consuming, and often fail to capture meaningful numbers of migrants. In addition, household surveys miss people living temporary housing and shelters, and commercial buildings as well as undocumented domestic workers living in their employer's residence. Targeted sampling or time-location sampling only work for populations that are geographically concentrated or are visible at common venues such as mosques, churches or other social organizations, health care facilities, temporary shelters and public squares (Watters & Biernacki, 1989; Muhib et al., 2001). This type of sampling often requires questionnaires to be short, post sampling weights to account for venue attendance frequency variation and a random selection of numerous sampling venues, as well as being prone to high non-response rates (Karon & Wenjert, 2012; McKenzie & Mistiaen, 2009). Snowball sampling, in which migrant individuals or households containing migrants are asked to provide referrals to other individuals or households, produce non-probability samples of unknown representativeness, making it difficult to generalize any conclusions reached from them (Bonnie, 1978).

In 1997, respondent driven sampling (RDS) was introduced as an alternative method to recruit and to provide generalizable estimates of hard-to-reach populations. Several aspects of RDS make is suitable for sampling migrant populations, especially those that are most hidden and are least likely to participate in surveys using other sampling strategies. This chapter describes how RDS works, providing examples of its suitability for hard-to-reach populations, and the functional and analytic assumptions upon which RDS is based.

How RDS works

Although RDS may at first appear to be a relatively easy-to-implement method, in reality many investigators have difficulty making sense of and meeting the strict assumptions and in properly following the implementation and analysis parameters of the method. Basically, RDS is a modified form of chain-referral sampling, whereby peers recruit their peers using coupons with unique code numbers (Heckathorn, 1997; 2002). As many migrant populations can identify others as members of their own group, relying on them to recruit other migrants is often a feasible strategy (Johnston & Malekinejad, 2014). Involved in the recruitment process are nominal incentives for survey participation and peer recruitment. Incentives, along with modified peer pressure, encourage people to enroll in the survey and to, in turn, influence their peers to enroll as well.

Recruitment is initiated with a small, diverse and influential group of "seeds" (eligible respondents) selected by the researchers. Each seed receives a set number of recruitment coupons to recruit his/her peers who then present the coupons at a fixed site to enroll in the survey. Eligible recruits who finish the survey process are also given a set number of coupons to recruit their peers. The recruited peers of seeds who enroll in the survey become wave one respondents, and the recruits of wave one respondents become wave two respondents. This process of recruitment continues through successive waves until the calculated sample size is reached. In the end, the waves produced by effective seeds make up recruitment chains of varying lengths. The goal is to acquire long recruitment chains made up of multiple waves. Figure 1.1, from the *sub-Saharan Africans in Morocco'* survey, (see Appendix I, for a thorough presentation of all surveys referred to in this volume) shows a recruitment chain made

12   Lisa G. Johnston

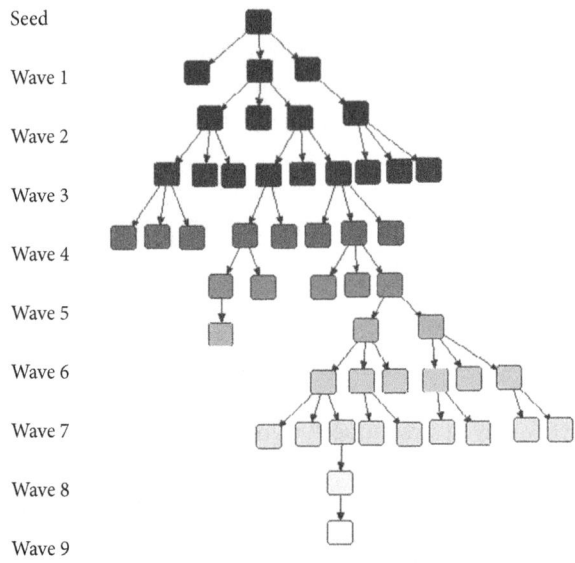

FIGURE 1.1   *Recruitment chain in RDS. Illustration of seed and waves*
Source: Authors simulation.

up of ten waves, including 50 respondents. To the left of the graphic is a side bar with the corresponding wave numbers.

The RDS sampling methodology minimizes several types of bias found in other chain-referral methods (Heckathorn, 1997; Johnston, 2013b). The use of "coupon quotas" (a set number of coupons, usually no more than three) for all respondents reduces the opportunity for those with larger network sizes to over-represent the sample with individuals who have characteristics similar to them. Allowing the use of a small number of coupons for each respondent also serves to lengthen recruitment chains, penetrating more deeply into the network and reducing sample dependence on the seeds. So even if the only available seeds are those who can be found at the only migrant social organization in town, long recruitment chains help ensure that individuals not associated with that migrant social organization are also included in the sample.

An incentive not only for participation, but also for recruitment, encourages traditionally hesitant individuals to participate in the survey, thereby reducing non-response. Additionally, the use of social pressure from a trusted peer, who is incentivized to recruit his or her peers, results in higher response rates. Finally, the use of coupons with unique numbers or codes allows respondents to remain anonymous, also reducing non-response bias. More hidden types of migrants, who normally would not participate in a survey, may feel more comfortable in enrolling knowing that researchers will not have any personal information that can be used to contact them later.

It is common that researchers gather data using RDS methods without understanding that RDS requires special analysis. Once data is collected using RDS methods, it must be analyzed to reduce biases by applying computational weights. Therefore, RDS must be considered both a sampling and an analysis method and every survey requires both methods in order to be called "RDS".

RDS assumptions

There are a number of assumptions in RDS, many of which are based on social science statistics (Heckathorn, 2007), and some of which are difficult to meet in real-world applications of RDS (Gile et al., 2014). The first assumption is that respondents know one another as members of the target population and recruitment ties are reciprocal. Basically, the target population must be socially networked and know (and be able to recruit) persons in that social network who also know them. For instance, just because someone is a "migrant" living in Madrid does not mean he/she is socially networked with other migrants. Migrants may be from different countries, speak different languages and have different customs, all of which would create barriers to forming social networks. However, migrants originating from a particular country living in Madrid may be socially networked and have reciprocal relationships with other migrants from that same originating country.

The second assumption is that there is sufficient cross-over between sub-groups and that networks are dense enough to sustain a chain-referral process. Although a population might be socially networked, there may be barriers within that network that would prevent members recruiting each other or there might be so few members that sampling cannot be sustained. Polish migrants living in Madrid, for example, might contain two distinct groups that never interact. The barrier here could, for instance, be between migrants with high and low socio-economic status. If these two distinct groups of Polish migrants do not recruit each other and they have distinctly different characteristics, then the final estimates from the survey will be unstable. Avoiding these types of barriers ensures that the entire network being sampled is one single network component rather than isolated clusters of distinct sub-groups. In addition, if there are too few Polish migrants in Madrid, the numbers will not allow the required sample.

The third assumption is that sampling occurs with replacement. Sampling with replacement requires that the sample size be small in relation to the population size. If the sample size required to create valid estimates is large and your population size is small, it is possible that achieving the target sample size would be impossible because respondents would not have a large enough pool of eligible individuals in their network to recruit. There is some debate about whether RDS can be a sampling with replacement method. On the surface, it is not, since a respondent is allowed to enroll only once (Gile & Handcock, 2010); however, the assumption is viewed more loosely if the sample size is small relative to the population size (Volz & Heckathorn, 2008).

The fourth assumption is that respondents are recruited from one's network at random. RDS assumes that respondents recruit as though they are choosing randomly from the pool of people they know who are eligible for recruitment. Non-random recruitment will not necessarily bias the RDS estimator as long as recruitment is not correlated with any variable important for estimation (Volz & Heckathorn, 2008). In practice, it is difficult to ensure random recruitment in RDS (Gile et al., 2014). The type or size of incentive, the interview venue or level of stigma and discrimination towards the population and the survey topic can influence whom respondents choose to recruit (Heckathorn, 2007).

The fifth assumption is that respondents can accurately report their personal network size, defined as the number of acquaintances, relatives and friends who can be considered members of the target population. The personal network size is measured by asking the respondent a series of questions that will lead to an estimate of the number of people they know, who would meet the eligibility criteria. The personal network size sets up the probability that an individual will be recruited and is used to calculate weights for data analysis. Essentially, those with larger network sizes have more paths that lead to them and are therefore more likely to be recruited than those with smaller networks, so they are assigned smaller weights. Respondents with small network sizes are assigned larger weights because they have fewer paths that lead to them.

The sixth assumption is that each respondent recruits a single peer. Ideally, RDS should allow for each person to recruit only one peer in order to avoid the bias of differential recruitment. However, to avoid recruitment chains from terminating, RDS allows for respondents to recruit slightly more, usually two or three. Many surveys now start with multiple coupons but reduce coupons once sampling is underway (Johnston et al., 2008) in order to reduce differential recruitment and improve variance (Goel & Salganik, 2009).

The seventh and final assumption is that a Markov chain model of recruitment is appropriate, resulting in a sample independent of seeds (Heckathorn, 2007). This is where the concepts of homophily and equilibrium are important. Homophily is the principle that contact between similar people occurs at a higher rate than between dissimilar people. This tendency often results in homogeneity of characteristics (most importantly socio-demographic characteristics) within respondent's personal networks (McPherson et al., 2001). Given that the seeds are purposefully selected, it is expected that the characteristics of the respondents in the initial waves of recruitment would be similar to those of the purposefully selected seeds. As a chain accumulates more waves, the bias from the seeds is reduced as new recruits enter the sample. *Equilibrium* is a diagnostic that measures the bias of the seeds. Equilibrium, also known as *convergence*, is the cumulative measure of proportions for a variable for each wave of the sample. For illustration, Figure 1.2 below shows the equilibrium estimates for the gender distribution in a sample. The vertical axis shows the percentage of males and females at each wave

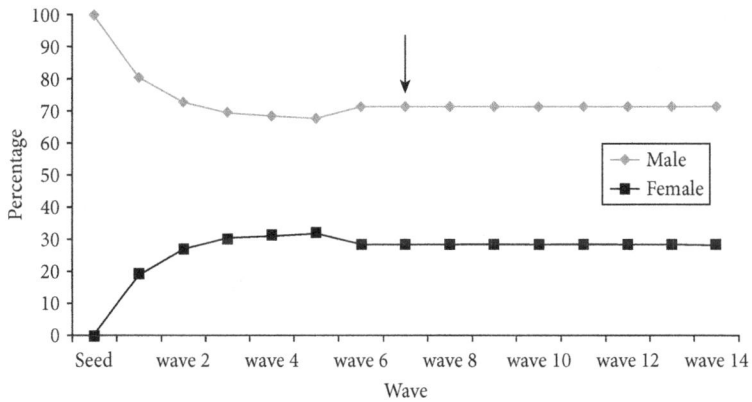

FIGURE 1.2 *Equilibrium for males and females*
Source: Survey of Anglophone sub-Saharan Africans in Morocco.

(horizontal axis). Equilibrium is attained when the proportions remain stable, within 2%, of the sample proportion. The proportions for this example appear to stabilize around wave 6 or 7 (see arrow), and remain stable until the final wave. The final sample comprises 71.5% males and 28.5% females.

Equilibrium is the point at which the sample is no longer biased by the seeds' characteristics, and is beginning to match the proportions of the characteristics of the network of the population being sampled.

Conclusion

RDS is a probability-based sampling method that offers solutions to numerous challenges found in many of the commonly used methods to sample migrants. If a traditional probability-based sampling design is easily available to sample a migrant population (i.e., a sampling frame exists or can be constructed, and there are no limitations in accessing migrants), it is usually best to stick with these methods rather than choose RDS. In addition, if the population is not socially networked, then some sampling method other than RDS may be more suitable. However, if considering the choice between a convenience sample, a probability-based sample that may suffer from numerous barriers to accessing migrants, and RDS, then sampling migrants using RDS is most likely the best option.

2
RDS and the Structure of Migrant Populations

Jon Horgen Friberg and Cindy Horst

Abstract: *This chapter addresses the appropriateness of RDS as a methodology for collecting survey data on migrant populations. What kind of problems do migration researchers encounter when using RDS to study migrant populations? What do researchers need to find out about the people they study, before starting an RDS survey? How do these issues relate to the structure of social relationships within the target population? The analysis is based on empirical examples drawn from a number of recent RDS studies that were carried out within migration research. We argue that in many cases, migrant populations are well suited to RDS research because the structure of these populations often corresponds well to some of the basic assumptions on which RDS is based. However, not every migrant population is suitable for an RDS study. Sometimes, researchers have to redefine their target population for the sampling process to work; at other times, RDS is simply not suitable, given the research problem at hand.*

Tyldum, Guri and Lisa G. Johnston, eds. *Applying Respondent Driven Sampling to Migrant Populations: Lessons from the Field*. Basingstoke: Palgrave Macmillan, 2014. DOI: 10.1057/9781137363619.0009.

Introduction

Respondent Driven Sampling (RDS), primarily developed within epidemiology and HIV research, has recently caught the interest of many migration researchers. But to what extent is RDS appropriate for collecting survey data in migration research? What kind of problems do migration researchers encounter when using RDS to study migrant populations? What do researchers have to know about the people they study before starting an RDS survey? In this chapter, we address these questions and how they relate to the structure of social relationships within migrant populations. We argue that in many cases, migrant populations are well suited for RDS because their structure often corresponds well to some of the basic assumptions on which RDS is based. However, that does not mean that this methodology is suitable for any given population of migrants. Sometimes, researchers have to redefine their target population for the sampling process to work; at other times, given a particular research problem, RDS is simply not suitable. When conducting RDS surveys, it is fundamental to learn how social relationships within the target population are structured, not only for asking the right questions (as is useful in any type of survey) but also for designing and conducting the sampling process itself. In other words, RDS researchers should go into the field and engage with their respondents in a different way than ordinary survey research usually requires.

Why RDS is well suited to studying migrant populations

RDS methodology is based on seven specific assumptions about the target population and the sampling process (Heckathorn, 2007) (See Chapter 1). This chapter addresses the first two assumptions, which state the specific conditions about the characteristics of the target population that must be met for the success of the RDS methodology. These assumptions are that respondents know each other as members of the target population (and recruitment ties are reciprocal) and that there is sufficient cross-over between sub-groups. The former and probably most essential assumption is that *respondents know one another as members of the target population*. This means that the population is linked by a pre-existing contact pattern and that these relationships are reciprocal: "I know you as a member of the target population and you know me as a member of the target population." If this criterion is not met, RDS is not appropriate for sampling: Although RDS is often used to study hidden populations, the members cannot be hidden from each other.

Let us examine the basic sociological concepts of *social categories* and *social groups*. Social categories consist of a collection of people who share certain similar characteristics but may not necessarily interact or recognize themselves or each other as members of the same group. For example, the elderly, millionaires, labor migrants, tax-cheats, refugees, high school students, disabled people and homeowners all constitute social categories. Although researchers may classify elderly in a social category based on objective criteria (i.e., those over the age of 64 years), not everyone in that category may identify themselves as elderly nor do they necessarily interact regularly with other elderly people. In much survey-based social science research, especially in commissioned and policy-oriented research, the population of interest is defined according to some kind of administrative category. It may be people who are subject to some particular set of policies or people who hold a residency permit based on certain criteria. In contrast, a social group is comprised of people who identify and interact with one another. Examples of social groups include: families, circles of friends, sports clubs and supporters, and ethnic and religious groups. One indication of whether a collection of people can be considered a group is if individuals who belong to it use the pronoun "we" when referring to the collective.

Members of a category can be distinguished from non-members in a precise way, for example, unemployed people under the age of 25 years. Members of a social category do not necessarily have any particular reason to be friends with each other, and if they are friends, they may not necessarily recognize each other as members of the same category. On the other hand, members of a social group interact and identify with each other; but social groups often have fuzzy edges: for example, it is difficult to say exactly where a circle of friends ends. And unlike social categories, which are defined objectively, the boundary of a social group often depends on an individual's perspective. For example, I might be able to tell who is and who is not part of my extended family, but my cousins will consider different (although overlapping) sets of people as belonging to *their* families. Therefore, the ideal target population in RDS studies is a social category of people that can be clearly defined by the researcher as well as by respondents that also has the characteristics of a social group in the sense that they identify and interact with each other.

How do such definitions apply to migrant populations? While some early immigration scholarship tended to see migrants as "uprooted" (Handlin, 1952), living in a context "divorced from a social setting" (Piore, 1979), more recent immigration scholarship shows that the process of migration is a highly network-driven phenomenon and newcomers are often linked by dense connections to other migrants who use their inside information to help their friends and relatives. Interpersonal relationships that link migrants, ex-migrants and non-migrants in countries of destination and countries of origin through bonds of kinship, friendship and common community origin and identity are recognized as one of the most important factors for explaining the process of international migration today (Massey et al., 1998; Arango, 2000; Epstein, 2008; Palloni et al., 2001). These networks are often among the most valuable assets held by migrants trying to adapt to a new country; they are a form of *social capital* that provides access to information about legal matters, job availability, housing opportunities, companionship and emotional support. Furthermore, ethnic or national origin tends to become a primary source of identity among immigrants in their host countries (although not necessarily for their children) by either ascription or self-identification. These networks and identities, which are so important to the daily lives of many migrants, are exactly the same networks that can help facilitate the RDS recruitment process. Ask any Somali refugee in the streets of Oslo or any Polish worker on a Lisbon construction site and they will likely have an extensive network of friends and acquaintances with the same national background who live in the same area and will have no problems identifying and distinguishing Polish or Somali immigrants from their other friends and acquaintances: a perfect starting point for RDS research. Nevertheless, there are several challenges and pitfalls that migration researchers need to be aware of when planning an RDS survey among migrants.

For example, RDS may not be suitable for migration research in sending areas. In a study of how migration systems evolve over time, the *THEMIS survey* researchers (see Appendix I, for a thorough presentation of all surveys referred to in this volume) intended to use RDS for sampling return migrants in Ukraine, Morocco and Brazil. Yet it turned out that these return migrants had no particular reason to know each other and it was difficult for both researchers and respondents to distinguish return migrants from non-migrants. In the receiving country, it is often very different.

But migrants are not only defined by strong local networks determined by nationality; they often also have strong transnational ties. Thus, the naturally occurring social group goes beyond national borders, even though the research focus may be restricted to a particular locality in the country of settlement. This factor does not necessarily constitute any problem in itself, but if the migrants also display a high degree of transnational mobility, the target population itself may become blurred. For example, parts of the Somali political and economic elite live as a "part-time diaspora" both in Europe and in the Horn of Africa, making best use of the opportunities of a transnational life (Hammond, 2011). A related concern occurs when the target population experiences cyclical or seasonal variation. For example, if targeting a population involved in seasonal migration (for example, to work in agricultural harvesting), the population will change over time so the timing of data collection is key. Because RDS data collection may often stretch over several months, surveys targeting seasonal migrants should be timed to minimize such changes within the data collection period, and the researcher must be aware that the analysis and results may depend on the season.

Target populations and naturally occurring social groups – common problems

As some migration researchers using RDS have learned the hard way, a good definition of the target population can mean the difference between a successful and an unsuccessful project (see, for example, Evans et al., 2011). Ideally, researchers want the best possible fit between their defined target population and some naturally occurring social group. This aim relates to the second assumption on which RDS is based. *Respondents must be linked by a network composed of a single component:* The entire population of interest must be interconnected through dense personal ties. We present examples here of how this has been achieved in concrete projects.

A bad fit between the target population and a social group can occur in two ways. First, the target population may include only a small section of some larger naturally occurring social group. Although social-network research on the so-called "small world phenomenon" has consistently shown that random people in different locations are connected to each other through surprisingly short chains of personal relationships (Milgram, 1967; Leskovec & Horvitz, 2008), this does not mean that any population necessarily constitutes a network composed of a single component. For RDS to work, all members have to be linked through other members of the same target population; it is not enough to be linked via outsiders. For example, sampling a specific age category within a larger immigrant group would not usually pose a problem, because young people tend to be directly linked to each other through personal relationships (i.e., not via older people); similarly, studying only women or only men within a migrant population, would probably not pose a problem, as most people engage in same-sex friendships and acquaintanceships. However, the sampling process is complicated when respondents are not directly linked to each other but via others. In the survey of *Central American Women in Houston* (see Appendix I), researchers were interested in sampling only recent immigrants (Montealegre et al., 2011), but soon realized that recent immigrants were not primarily connected to each other directly but via more established immigrants, who constituted the "glue" in the immigrant community; the researchers redefined their population to include all Central American Women in the area.

Second, a bad fit between a target population and a naturally occurring social group may also occur when the target population encompasses two or more groups that are weakly linked to each other or not linked at all. For example, when initiating the *2006 Polonia in Oslo* survey, the researchers realized that their target population, Polish migrants, consisted of two social groups that have little contact with each other: a small group of Polish activists, academics and dissidents who fled to Norway and were taken in as political refugees after the *Solidarnoc* uprising in the early 1980s, and the large and rapidly growing number of relatively low-skilled labor migrants who arrived after Poland's accession to the EU in 2004. Belonging to different social classes, age groups, cultural segments and migration waves, these two groups were hardly linked by personal relationships and they did not identify much with each other. The solution was to exclude the earlier refugees from the target population by introducing the criteria that respondents must have arrived in Norway after 1991 (after the last refugee was settled and before the arrival of the post-communism labor migrants).

These two cases seem to contradict one another. What appeared to be the problem in the *Central American Women in Houston* survey – introducing a cut-off point in terms of when respondents arrived – was the solution in the *2006 Polonia in Oslo* survey. On closer inspection, however, both illustrate the need to define or redefine the target population to get a better fit with the network structure of a naturally occurring social group. In the following section, we will discuss different types of network structuring that may affect the sampling process, using the terms *bottlenecks* and *clustering*.

Bottlenecks and clustering

In RDS terminology, parts of a population that are more densely connected than other parts are referred to as *clusters* and few connections between particular sub-groups are referred to as *bottlenecks*. A cluster refers to the clustering of personal links within sub-groups; a bottleneck is the absence of personal links between different sub-groups within the target population. When migrants come from one particular country, several kinds of divisions among them may lead to clusters and bottlenecks in the migrant population.

Members of the target population may belong to different *migration waves* as in the previous *2006 Polonia in Oslo* survey example. Another case was found in the *THEMIS-UK* survey: The Ukrainian community in the United Kingdom consists of substantial numbers of both post World War II immigrants and post-communism immigrants, separated by 40 to 50 years of residence in the United Kingdom. Relatedly, they may belong to different social classes. For example, highly skilled professional migrants or students may have little contact with low-skilled labor migrants from the same country. They may come from different regions or towns, leading to the formation of geographically determined sub-groups. They may belong to different ethnic and linguistic groups who hardly interact, as in the case of Turkish Kurds and other Turkish immigrants in many European countries. Or they may belong to different political factions or parties involved in conflict that, in many cases, coincide with ethnic and linguistic differences (for example, Tamil and Singhalese immigrants have little contact with each other in the host country even though they both come from Sri Lanka; the same is true for Shia and Sunni Iraqis or Bosnian Muslims and Bosnian Serbs). Finally, in some groups with strict norms regulating contact between men and women, gender may pose a bottleneck: Although men and women are linked by personal ties in all populations, in some populations, gender norms may prevent people from recruiting across the gender divide. To avoid bottlenecks in the sample, researchers not only have to take into account the population structure, but also the structure of recruitment. Such issues will be addressed in the chapters on survey planning and implementation and on the training of interviewers and seeds (see Chapters 4 and 6).

Bottlenecks may vary in different social settings. Usually, language and nationality will serve as important dividing lines, but sometimes other traits may prove more significant. For example, in a study of "Foreign Migrants in Ukraine", the inclusion of Russian and Belarusian students in the same sample posed no problems. However, the division between students of different faculties and science disciplines turned out to be a major bottleneck. Similarly, the bottleneck between students and workers proved significant in the study of Ukrainian, Russian and Belarusian migrants in the *Migrants in Warsaw* survey.

Most populations will display some clustering and bottlenecks for particular variables, and the RDS estimators are able to adjust for this to a certain point (see Chapter 7). However, clustering and bottlenecks can add variance to estimates. In the example of the *2006 Polonia in Oslo* survey, the researchers decided that the bottleneck between the refugees and the labor migrants was too great to overcome within one survey; subsequently, they decided to treat them as separate populations and target only the largest and most recent one. In most cases, RDS surveys must deal with some level of clustering and bottlenecks. Being aware of them lets the researcher handle possible distortions. It is important to ask the right questions to identify and monitor bottlenecks, and the researchers can help reduce variance by ensuring that different sub-groups of the population are represented among the original seeds and that respondents try to recruit across the sub-groups (see Chapters 4, 5 and 6).

Getting to know the study population

We have illustrated the importance of knowing a target population's social structure before RDS researchers begin a survey. RDS is often used to study hard-to-reach populations, where little information is available. Therefore, researchers have found it essential to collect some information before initiating data collection (Johnston et al., 2010; Johnston, 2013b). We have argued how prior information is crucial for defining an eligible target population and for monitoring bottlenecks in the sampling process.

The need for prior knowledge on the social structure of a migrant population under study makes RDS perfectly suited for a mixed methods approach. In the *THEMIS* surveys, RDS was the third step of the research design; following an exploratory "scoping" study with six migrant groups in four European countries and an in-depth qualitative study in the country of settlement and origin of the three selected migrant populations, their family members, and returnees. The qualitative studies provided in-depth knowledge of the social structure of the three groups, which was useful in designing the RDS phase. However, RDS studies are not necessarily conducted as part of a mixed methods approach, but getting the necessary information about population structure enables a sound design of the RDS study.

Before defining their target population, therefore, RDS researchers need to engage in some type of initial qualitative inquiry – often called *formative assessment* or a *scoping study* – to ensure that the target population corresponds to a naturally occurring social group whose members are directly linked to each other. The inquiry should prevent the selection of a group that is divided into separate rarely interacting sub-groups, and it can be used for identifying potential bottlenecks. An initial qualitative inquiry typically consists of informal open-ended talks with experts, key informants and members of the target population. Some researchers use individual interviews; others prefer to organize a focus group. A good starting point could be approaching those familiar with the particular group, such as community leaders, churches and congregations, organizations that represent group members, political activists, or academic or other experts. However, since elites may not always be as well informed about the lives of ordinary people as they think they are, it is usually wise to also include interviews with regular members of the target population.

Migration researchers using RDS should be aware of ethical issues that relate to migrants: Migrants are often vulnerable both economically and legally, they might distrust or fear authorities, and are sometimes stigmatized in their host countries. This calls for extra consideration in how researchers interact with the people they study, how they obtain informed consent, and in storing and protecting sensitive data and in how their findings may be interpreted and used in the media and public discourse. Most of these issues are no different from ethical issues that all migration researchers (and social scientists in general) have to deal with. However, because RDS is often used to study hidden and hard-to-reach populations where no sampling frame exists (such as recently arrived, unregistered or irregular migrants), these issues are often particularly pertinent. Furthermore, certain ethical considerations particularly apply to RDS research, including the use of economic incentives and chain-referral recruitment (see Chapter 5).

Conclusion

RDS is an appropriate and useful method for collecting survey data in many types of migrant populations that are difficult to sample through traditional sampling techniques, because the structure of migrant populations corresponds well with the assumptions upon which RDS is based. However, we stress that this cannot be assumed but needs to be explored: It is crucial to engage in an initial formative assessment to get to know the social structure of the migrant community being researched, before designing the RDS study. Two common concerns are that the selected target population is only a part of a larger social group, such as when one wishes to study recent migrants only, or that the selected target population consists of sub-groups that barely interact, as in the case of Poles in Norway or Ukrainians in the United Kingdom.

3
Measuring Personal Network Size in RDS

Lisa G. Johnston, Leila Rodriguez and Joanna Napierala

▶ **Abstract:** *This chapter presents the importance of the Personal Network Size (PNS) variable, and its collection and use. The PNS variable is one of only two required pieces of data needed from an RDS survey, and is crucial for meeting the assumptions necessary for statistical analysis. Measuring an individual's PNS requires that they respond to a question, or series of questions, on the number of eligible people the respondent knows, and who also know the respondent. The response must be a number greater than zero. Obtaining this measure is not as simple as it may first appear. One challenge involves delimiting, in a culturally compatible manner, what it means to "know" someone. Another challenge is to train staff to work with respondents so that the latter can accurately estimate this number. Breaking down the question into several parts and using probing techniques have proved to be useful in some surveys.*

Tyldum, Guri and Lisa G. Johnston, eds. *Applying Respondent Driven Sampling to Migrant Populations: Lessons from the Field*. Basingstoke: Palgrave Macmillan, 2014. DOI: 10.1057/9781137363619.0010.

Introduction

Respondent Driven Sampling (RDS) is employed in surveys to collect all sorts of information, but only two pieces of information are mandatory to collect: the connections between recruits and recruiters, which we track through the use of coupons, and the respondent's personal network size (PNS). This chapter focuses on the PNS indicator that measures how many individuals from the target population respondents know, who also know them (Heckathorn, 1997). Each respondent's PNS is measured through an open-ended question, or series of questions. Constructing the PNS question(s) and collecting accurate responses from respondents is not straightforward. Given the challenges to eliciting responses to the question(s), researchers use prompting techniques, and often place the question(s) near the beginning of the survey. In this chapter we use experiences from past RDS surveys to help demonstrate how to accurately obtain PNS data, as well as providing answers to several common questions, including: What is the PNS question(s) and why do we need it? How do we construct the PNS question(s)? And, what are the challenges in eliciting accurate responses for the PNS question(s)?

The PNS variable and why we need it

The PNS variable measures the number of eligible persons in each participant's personal network (Heckathorn, 1997). Each respondent's PNS is measured with the PNS question(s), which provides the variable to assess sample bias and weight data during analysis. The assessment of these two factors shows the non-random manner in which respondents are sampled (see Chapter 7). Measuring each participant's PNS is important since it constructs the sampling frame, which is the probability of selection based on the size of each respondent's PNS (versus the probability of selection based on population size as is used in most traditional probability sampling methods). During analysis, differential PNS measurements are adjusted to prevent over and under representation of certain sub-populations during analysis. It is essential for the PNS question(s) to be constructed properly so that responses are as accurate as possible.

Constructing the personal network size question(s)

Four elements are needed when constructing the PNS question(s): a clear definition of the target population; a shared understanding or explicit definition of what it is to "know" someone; a geographic boundary for the survey; and, a clearly defined reference period during which the respondent has come into contact with the reported peers.

Clear definition of the target population

The wording of the PNS question(s) is developed directly from the eligibility criteria of the population being studied. For instance, in a survey of individuals aged 18 years or older, who originate from francophone sub-Saharan African countries, live and/or work in Rabat, and have resided for at least three months in Morocco, the PNS question could be phrased: How many individuals do you know (who also know you), that are aged 18 or older, are from francophone sub-Saharan African countries, and have been living and/or working in Rabat, have resided for 3 months in Morocco and who you have seen in the past two weeks? In practice, however, the PNS question is often split into several sub-questions.

For RDS, it is essential that the target population forms a social network, that the definition of the network is clear, and people are able to identify individuals who fulfill that definition (Johnston, 2013b). For example, if researchers are only interested in "recent" migrants, or "undocumented" migrants, individuals need to be able to identify these traits in their migrant peers. If they do not know this information about their peers, a situation known as *transmission error*, they may not be able to accurately report on the number of individuals they know in that specific population (McCormick et al., 2010). In situations where populations are not networked, and important traits within their networks cannot be identified, modifications on the definition of the population should be considered. Specific examples of such situations were encountered in the *Polonia in Oslo* survey (see Appendix I, for a thorough presentation of all surveys referred to in this volume), whereby the 22-year gap between inflows of Polish migrants in Norway resulted in two separate networks (one of Poles arriving before 1991 and one of Poles arriving after 1991) making it necessary for researchers to focus the eligibility to Poles arriving after 1991. Among Russians in Costa Rica, migrants who arrived as Soviets in the 1970s were not networked with those arriving from Russia after the breakup of the Soviet Union (Rodriguez, 2005).

The meaning of "knowing" someone

We all have different ideas of what it means "to know" someone. To some individuals, "knowing" someone is to be able to provide their first and last name and address, phone number or email, and seeing them regularly. These are considered strong ties. To other individuals, "knowing" someone is to recognize them by sight or be familiar with them by only their first name or nickname, and to encounter them infrequently. These are considered weak ties. For RDS, we wish respondents to recruit from both weak and strong ties to ensure that a broad spectrum of population members are included in the survey and that recruitment extends into potential bottlenecks (i.e., visible individuals may be linked to more hidden sub-populations by weak ties). In the survey-planning phase, it is essential to understand how target population members define what it means to "know" someone and to get an overview of the alternative ways of phrasing this, especially when developing the PNS question(s) in foreign languages that may have more than one verb for "know". Understanding these points can ensure that both strong and weak ties are included as part of someone's PNS. Inherent in knowing someone is the idea of reciprocity, which means that recruits and recruiters know each other. This factor is usually assessed by including in the PNS question: "Do you know them and do they know you?" or "Do you know their name and do they know yours?"

In some RDS surveys, especially in areas where social media are prevalent, knowing someone could be measured by the number of people they have "seen" on Facebook, Twitter, or other social media venues. Other RDS surveys have asked about how many people a respondent knows through email or texting. Keep in mind that measuring PNS as the number of people they interacted with through social media or other technologies in the last two weeks may be inversely correlated to the number of people they had face-to-face contact with during the same period. Furthermore, the idea behind knowing each respondent's PNS is to assess the probability of being recruited and, in most surveys, with the exception of internet-based RDS surveys (Wejnert, 2009), it is usually necessary for individuals to meet face-to-face in order to pass on a coupon.

Geographic boundary

Estimating the number of people we know personally is not straightforward, and can become even more complicated with the dynamics of migration. When constructing the PNS question(s), a reasonable geographic boundary is necessary in order to generalize findings. For example, if you are conducting a study of Polish migrants in London, and simply ask "How many Polish migrants do you know?" respondents may include all Poles they know who live both within and outside London. In this scenario, the PNS variable includes individuals who never had any probability of being recruited into the study (i.e., those living outside of London). This geographic boundary is often part of the definition of eligibility, as shown in the example above.

Time frame in which the respondent has seen their peers

Specifying a time, during which the respondent has seen their peers, allows for an accurate PNS measure that reflects the number of people the respondents are likely to encounter and could recruit for the survey. For instance, an individual, who knows the entire ethnic community from meeting its members on national holidays or at other events, may only meet with a few persons from this group on a day-to-day basis. It is the individuals they meet or see in day-to-day interactions that are likely to recruit them for the survey, and it is the number of these we wish to estimate. The period of time is usually added to the PNS question as, "and have you seen them in the past two weeks?" or some other appropriate period. It does not matter so much if your PNS question uses a two- or six-week period throughout the survey, as the effect of increasing and decreasing this time is not likely to produce biases between population groups. However, the shorter the estimation period, the better quality data can be expected, as you reduce recall bias. In original RDS surveys of people who inject drugs, the time period for when a respondent saw their peers was six months (Heckathorn, 1997), which may be too long a time for many people, especially those with large networks, to accurately recall. One recommendation is to use a time period reflective of the amount of time needed for recruiters to pass on coupons and recruits to redeem them – this information is easily gathered through formative assessment (Gile et al., 2014).

As mentioned above, the PNS question can comprise one question or a series of questions. Many surveys have found that having just one question with so many elements may be too difficult for some people to answer. These surveys have divided the question into a series of questions beginning with a broad question and narrowing it down to the final question that will ultimately be used in the analysis of RDS data

(Johnston et al., 2008; Johnston, 2013b). Each question under the first question forms a sub-set of the previous question.

Figure 3.1 provides an example of the PNS question from the "Central American Women in Houston" survey, and is divided into four separate questions, with the first question being a general one on the number of people the respondent knows that meet the eligibility criteria. If there are several parameters for eligibility, researchers may decide to use them all to make two separate questions (for instance, in the survey of *sub-Saharan Africans in Morocco*, eligibility included age, country of origin, residence and length of residence parameters (see above)). Having the first question as a broad one, allows respondents to think more easily about the population. The response to the first question, therefore, does not need to be precise. The second question in Figure 3.1, which is often built into the first question, defines what it is to "know" someone, to ensure that everyone has the same understanding of this concept, and that relationships are reciprocal. The third question is a sub-set of the first and second questions and the response to this question is expected to become more precise. Finally, the fourth question sets up the period

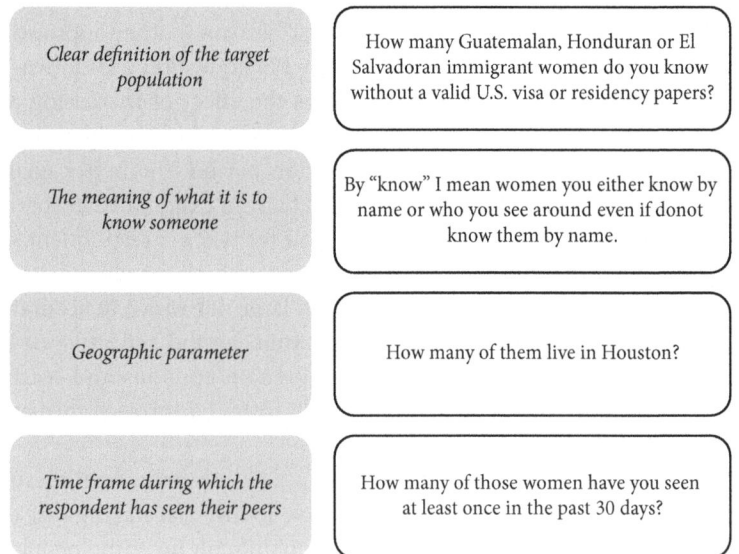

FIGURE 3.1   *Personal network size question*
Source: Survey of Central American Women in Houston.

during which the respondent has seen members of his or her personal network. The response to this question, as sub-set of the previous question, should be as precise as possible since it will be used in analysis.

Measuring PNS

Once you have constructed a clear and understandable PNS question(s), there are numerous other challenges to consider. Below are several examples of challenges and the ways they have been overcome in different surveys of migrant populations.

Eliciting PNS by sub-group

Some migrant populations are extremely large, and for migrants whose entire social world could consist of that migrant community, you may be asking them to recall a very high number of individuals. One method to help respondents recall the number of people in their PNS is to divide the network into sub-groups. For RDS, sub-groups of the target population could be classified by occupation, nature of the relationship (e.g., family, close friends, acquaintances), and places where sub-groups are encountered (e.g., work place, market, church or mosque (Bernard et al., 1991; McCarty et al., 2001). Eliciting the PNS question(s) in this manner requires the researcher to know something about the population. For instance, in the *Nigerians in NYC* survey, formative assessment was conducted to identify important sub-groups. The important sub-groups for this population consisted of ethnicity (Nigeria is composed of about 250 ethnic groups, but only a handful are readily represented among the New York City migrant population), social and communal institutions, such as the church or mosque, work place, and the ethnic or hometown associations in which Nigerian immigrants frequently participated. Once this exercise was completed, respondents were asked to report on one final number as their PNS. The numbers by sub-group were not simply added up, because some of the categories overlapped. The point of the exercise was to get respondents to consider all the types of individuals they may know who constitute part of their PNS.

Training staff

Extra care is needed when training staff on how to elicit information about the PNS and how to use the PNS to explain the recruitment process. Staff should be trained how to prompt when respondents have difficulty answering the PNS question(s) or when they claim not to know anyone in the target population. Measurement error in the PNS can introduce biased estimates, especially for variables for which the percentage is low (Gile et al., 2014).

In addition, the PNS response is essential for helping respondents think about whom they should recruit, and for making them recruit as broadly as possible (see Chapter 4). In explaining the recruitment process to recruiters, it is useful to ask them to recall the people in their personal network they said they knew (it is also useful to have that number available in order to remind them) and then to ask if they can try and recruit from those.

PNS of zero, outliers and coarsened data

Any respondent in an RDS survey will know someone in the target population, having already been recruited by someone they know; so all responses must be greater than zero. If a respondent insists on not knowing anyone in the target population or if some of these values are missing, you will have to impute new values since missing values for PNS variable will bias the estimators. One way to deal with this is to impute the PNS variable for any respondent who has a zero PNS as 1 (0+1) if they recruited no one, 2 (0+2) if they recruited one person, 3 (0+3) if they recruited two people, and so on. However, responses of zero or missing data should not be allowed in any RDS surveys, and interviewers can be trained to implement this.

If respondents report an exceedingly high PNS, some adjustment may also be necessary, as this will have the effect of reducing these respondents' weight in the analysis. As the values of the PNS increase, individuals often start to round the values. For instance, if someone says they know 80 individuals, they tend to *heap* or *coarsen* data, meaning that they know between 70 and 90 individuals. In other words, the person does not know the exact number, so gives an estimation. One way to control this is to have a short recall period in the PNS question. Another solution is to consider running a histogram of the PNS values to identify *outliers*. Figure 3.2 displays the PNS values of sub-Saharan Africans in Morocco by wave, starting from wave 0 (the seed) up to wave 7. The horizontal lines across each wave, display the mean PNS value. The larger the black dots, the higher the number of respondents with the same PNS. In this

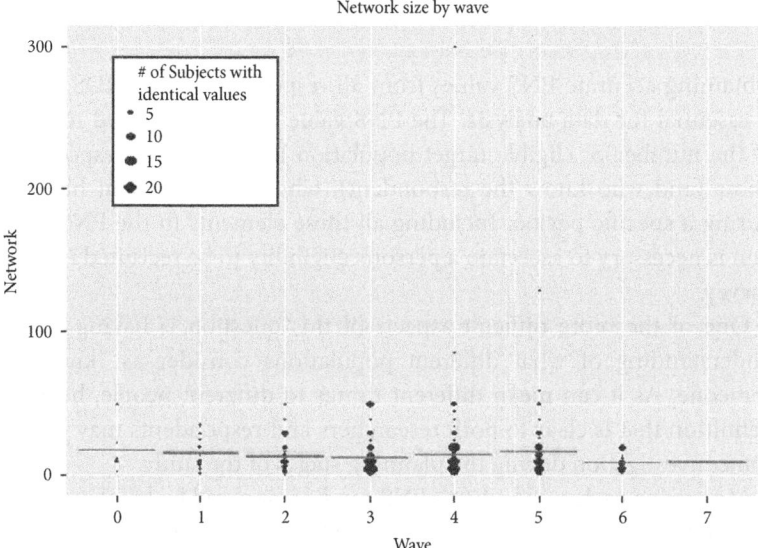

FIGURE 3.2 *Histogram of personal network size by wave*
Source: Survey of sub-Saharan Africans in Morocco (Created in RDS Analyst software (www.hpmrg.org).

example, the mean PNS values are fairly stable across waves; however, a few PNS values are larger than the others. Look at wave 4, which shows a respondent with a PNS value of 300, and wave 5, which shows PNS values of 250 and 125. If there are outliers, consider putting them to the value at the 75th percentile or to a reasonably lower value. In this case, bringing all PNS values down to 100 would seem plausible.

Temporal impacts

Some groups of migrants (i.e., circular or seasonal) are especially mobile, dividing their lives between two countries. Hence, when planning an RDS survey, it is worth remembering the high mobility in a few specific periods and events during the year (e.g., Christmas, Ramadan or other holidays, seasonal changes in demand for foreign labor, etc.) when the networks of migrants are different from the networks in the rest of the year. Asking the PNS question during periods of high mobility can result in inaccurate PNS measurements and cause instability to the estimators.

Conclusion

Obtaining accurate PNS values from all respondents in an RDS survey is essential for data analysis. The PNS value is a self-reported measure of the number of eligible target population members that respondents know (and who know the respondent), who the respondent has seen during a specific period. Including all these elements in the PNS question is necessary to ascertain a person's eligibility to be recruited into the survey.

One of the more difficult aspects of this question is having a clear understanding of what different populations consider as "knowing" someone. As it can mean different things to different people, having a definition that is clear to both researchers and respondents may require some investigation during the planning stages of the study.

Measuring each respondent's PNS can be improved by breaking down the PNS question into several sub-questions, eliciting PNS values by sub-groups, providing adequate training to staff, probing PNS response values of zero or values that are exceedingly large, and by planning surveys that avoid excessive in and out migration.

4
Initiation of the RDS Recruitment Process: Seed Selection and Role

Agnieszka Kubal, Inna Shvab and Anna Wojtynska

Abstract: *This chapter addresses the initiation of RDS, and examines the selection and role of the seeds: the first respondents in the surveys using RDS methodology. It is organized around three main questions: Who are the seeds? How do they work? Why are they important? We address these questions paying particular attention to the strategic selection of seeds, their training and role in the day-to-day sampling process. Therefore, although the selection of seeds takes place at the beginning of a survey, this process needs careful consideration, as it is likely affect data collection and analysis.*

Tyldum, Guri and Lisa G. Johnston, eds. *Applying Respondent Driven Sampling to Migrant Populations: Lessons from the Field.* Basingstoke: Palgrave Macmillan, 2014. DOI: 10.1057/9781137363619.0011.

Introduction

Researchers, who have used Respondent Driven Sampling (RDS) often, in hindsight, asked themselves: "what do we wish we had known before embarking on the first RDS survey"? The response is often: how to initiate the RDS recruitment process – how to select the initial respondents and make them work as they were supposed to; in other words: *to plant the seeds in such a way that they would grow and bear fruit.*

This chapter focuses on initiating the RDS recruitment process, and examines the role of the seeds: the first respondents in surveys using RDS methodology. We have organized the chapter around three main questions: Who are the seeds? How do they work? And, why are they important? In the following sections, we address these questions, paying particular attention to the strategic selection of seeds, their training and their role in the day-to-day sampling process, as well as the potential biases introduced by the seeds to the final sample.

Strategic selection of seeds

The seeds in RDS are the starting points for peer-to-peer recruitment in a networked sample. As these first respondents are purposefully (and not randomly) recruited by the survey researchers, seeds introduce bias into the sample. To reduce this bias, the seeds are expected to initiate long recruitment chains of multiple waves of recruits to ensure that the sample reaches equilibrium: the point at which the sample is no longer influenced by the initial (biased) seeds. Figure 4.1 demonstrates how the characteristics of a seed can dominate the characteristics of the initial waves of a survey. The initial respondents invariably share the characteristics of the seed (e.g., being married) since individuals tend to recruit others similar to themselves (unless they were explicitly asked not to). In this example, the first unmarried person in the chain shows up in the wave 4. In the subsequent waves, more unmarried people appear, resulting in a mix of married and unmarried individuals.

Although RDS theory is based on an assumption that the seeds need not be diverse with respect to characteristics, having diverse seeds will help attain equilibrium more quickly (Heckathorn, 1997). Heckathorn based this observation on a strong assumption that the members of the survey population are well networked among themselves, and that within the

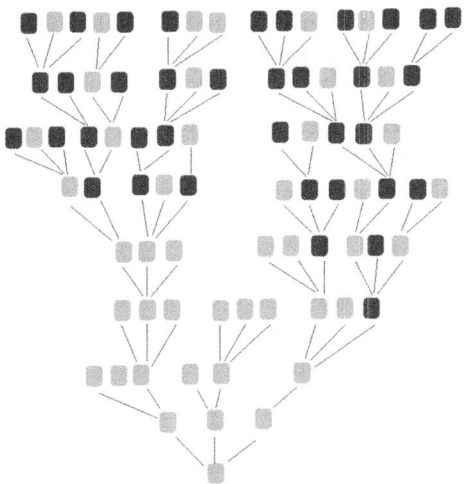

FIGURE 4.1 *Initial seed bias in recruitment of married and unmarried respondents*
Legend: grey – married, black – unmarried
Source: Authors' simulation.

target population, sub-groups according to, for instance, gender, ethnicity, education, and socio-economic status will not be completely isolated from each other. These assumptions are known to hold true for RDS surveys with hard-to-reach homogenous groups that represent a relatively small proportion of the overall population (Heckathorn, 2007). Not surprisingly, the most prominent examples of the groups where RDS has been successfully applied are among populations at high risk of HIV, such as people who inject drugs, sex workers, and men who have sex with men (Montealegre et al., 2013; Johnston et al., 2008; Malekinejad et al., 2008).

These theoretical assumptions might, however, be revisited when applying RDS to migrant populations. Chapter 2 discusses the social fragmentation and differentiation of migrant populations and introduces the concept of a bottleneck: the limited connections between sub-groups and their influence on the patterns of recruitment. Several studies have demonstrated that migrants are anything but a homogenous group of foreign-born people (Eckstein, 2009; Kubal & Dekker, 2011; Wimmer & Glick-Schiller, 2002). They may have arrived in the host country at different times, some of them may be highly skilled and others low skilled; some may be religious – others not; some will be legal residents – others, undocumented. Knowing the target population (its socio-economic parameters, potential bottlenecks, models of networking) in advance may significantly help in achieving a successful selection of seeds.

Formative assessment in RDS (see Chapter 6) is focused on getting to know the target population (alongside testing the survey tools). It may include in-depth interviews, focus groups, key informant interviews, document analysis, mapping, and observations of the target population (Johnston, 2011; 2013b; Johnston et al., 2010). Many of the RDS surveys among migrants discussed in this book were preceded by an in-depth study of the target population focusing on migration history and the dynamics of migration processes, sub-groups within the migrant population, the evolution of communities in the destination areas, return trends, employment niches, and analysis of secondary data, etc.

This more or less extensive formative assessment enabled researchers to find out which migrant sub-populations were likely to be most fragmented and which might respond positively to the RDS methodology. Above all, the formative assessment helped researchers to locate their seeds and decide on the optimal number of seeds (often corresponding to the number of most significant bottlenecks). For example, in the *THEMIS-Netherlands* survey, the population was characterized by homophily (see Chapter 7), as unemployed Moroccans were connected mainly to other unemployed Moroccans. Therefore, it was advisable to choose seeds representing different characteristics, and particularly people with different employment status who were also well connected within the migrant community, across various sub-groups.

Of course, prior knowledge of the target populations (be it migrant or non-migrant) can never be perfect. Bottlenecks around formal community institutions with which migrants are involved might be relatively easy to spot, but bottlenecks around informal ones – such as shops that people frequent – may be more impervious to direct observation (Goel and Salganik, 2009). For example, in the 2006 *Polonia in Oslo* survey, one large construction site proved to be a significant bottleneck as workers focused on recruiting other workers from that site.

Given the uncertainty about the links between the sub-groups identified during formative assessment, and potential (still unidentified) bottlenecks, it is important to spread the seeds between sub-groups and to monitor the linkages between them and their referrals. In *the Polonia in Reykjavik* survey, researchers decided to locate seeds representing the various occupations most common among migrants, which also were connected to their time of arrival. Knowing of the development of migration from Poland to Iceland, researchers expected that there might be fewer contacts between earlier waves of migrants arriving to work in the fishing industry (since the 1970s) and more recent waves of construction workers (since 2004). However, it turned out in the course of the survey, that there was sufficient interaction between the groups, and Polish construction workers ended up recruiting Polish migrants from the fishing industry and vice-versa. The experiences of earlier RDS research reiterate the importance of a strategic selection of the initial respondents to have all (or the majority of) the features the researchers wish to be represented in the sample.

It is, however, not always the case that having diverse seeds with large social networks will ensure that sampling results in a diverse mix of the target population. In the survey among *sub-Saharan Africans in Morocco* researchers achieved this by selecting seeds, not only with diverse characteristics, but based on their ability to recruit people with diverse characteristics. For instance, it was considered likely that sub-Saharan Africans in Morocco would form bottlenecks across countries of origin (i.e., migrants from Mali would only select migrants from Mali). The researchers therefore selected as seeds persons who knew individuals of several different nationalities. In addition, they created a grid of major sub-groups including country of origin, sex, age, and employment status, working with each seed to plan whom to recruit in order to ensure a diverse mix of characteristics early on in the sampling. Taking such steps at the start-up of the survey, makes the need for adjustments less likely later on. Lack of such deliberations could result in biases if the exclusion of certain important sub-groups is discovered in the middle of data collection (Johnston, 2013b).

Identifying seeds

How does the strategic selection of seeds look in practice? Once researchers have identified the kinds of seeds they need, the question becomes, where and how to find them. Formative assessment is useful for identifying potential seeds for populations in which the characteristics are unknown. In particular, while looking for seeds, researchers may consider the following auxiliary questions (however, the list is not exhaustive of the different conditions under which seeds can be found):

▸ Are there associations of migrants in the city/country, where the survey is conducted? Can these associations help identify seeds?

- Is there a religious center where migrants gather in the city/country? Does a religious leader agree to cooperate in identifying and locating seeds (to accurately describe the survey, to allow posting a notice about the survey, etc.)? What days of the week are most suitable for setting up contacts with migrants who observe a religion?
- Are there any international, non-governmental or religious organizations (e.g., trade unions, the Red Cross or Caritas) that work with migrants in the city/country? Can they help to locate seeds?
- Where do migrants spend their free time? Are there coffee houses owned by migrants, clubs, ethnic food restaurants, colleges, community centers, etc. in the city/country?
- Is there an embassy or a consulate in the city/country? Is it appropriate to search for seeds near the embassy? Are there periods when migrants are more likely to be at the embassy?
- Have other types of research been conducted among migrants in the city/country? Can other researchers help identify seeds?

In practice, many RDS surveys with migrant populations use a combination of the above techniques: "planting" the seeds simultaneously in a migrant organization, in a church, and in a business establishment pertaining to an ethnic niche of the labor market dominated by migrants, in order to facilitate linkages between various sub-groups.

In the *THEMIS-UK* survey of Brazilians, Ukrainians and Moroccans, four separate organizations were identified for these three populations: "Casa do Brasil", the Ukrainian Institute and two Moroccan associations –a female community center "Al-Hasaniya" and "Al-Noor" (near a Moroccan school). Of these four places where seeds were selected, only the Ukrainian Institute and Al-Hasaniya yielded long and diverse recruitment chains with RDS. The Ukrainian Institute is an organization that was originally established by the older generation of Ukrainian diaspora that arrived in the United Kingdom after the Second World War but has recently started attracting more economic, labor migrants by offering affordable English language courses. This venue was known and frequented by many different members of the Ukrainian target population, as the repertoire of activities it offered had evidently bridged the interests of the various Ukrainian migrant sub-groups. Al-Hasaniya, on the other hand, was the only Moroccan community center for women in London, targeting all ages: it had been organizing lunches for elderly women, offering family and legal advice for women, and had a nursery for children. The majority of the Moroccan female interviewees in the study, therefore, came through this venue. Comparing the results of the project to previous studies of Moroccans in London (Cherti, 2008) the researchers considered the ability to capture this rather "invisible" group of Moroccan female migrants as a great advantage of RDS (Bakewell et al., 2012). Religious institutions were yet another powerful center of communication among migrants. In the project researching Ukrainian labor migrants in Greece, the Ukrainian Greek-Catholic Church was useful for identifying seeds to participate in the survey. Lastly, ethnic business niches have also proven to be a good place to start. In the same survey of Ukrainians in Greece, aside from the church community, a seed was also "planted" in a transport agency specializing in transporting people (mainly migrant workers) between Greece and Western Ukraine.

Yet, it is important to keep in mind that selecting the seeds only from migrant organizations might introduce some bias to the sample, as places of interaction can impact who is interviewed (active community members, religious people, etc.). Researchers should take into account that usually, a significant proportion of migrants are *not* associated with any ethnic, community or religious organizations.

What other types of characteristics should seeds have? For recruitment to proceed smoothly, it helps if seeds are trusted, respected and well known in their community. Some researchers say that the level of respect and influence of a seed in a group is closely related to others wanting to participate in the survey – and thus helpful to reach across various bottlenecks within the population. However, other researchers found that community leaders did not necessarily work well as initial seeds (see Chapter 6). The soft skills of a seed – their ability to convince others to take part in the study – are nonetheless very important. In certain cases, migrants may be reluctant to participate in surveys due to their vulnerable position in the society. In such cases, knowledge of a person who is respected within the group and can vouch for the survey organizers, may play a key role in convincing group members to take part. For example, in the *THEMIS-UK* survey of Ukrainians, the most "productive" seed turned out to be a local community organizer, who was extremely active within the community, acting as an interpreter, social worker and editing a weekly newsletter with important information about events and services for Ukrainian migrants. He would turn out to be a "super-seed" – giving rise to long recruitment chains made up of people with diverse backgrounds and characteristics.

However, even a random selection of seeds may bring positive results. For example in the 2006 *Polonia in Oslo* survey, researchers selected as a seed a Polish construction worker they met at the work site. Although it was not certain whether he had good communication skills or whether he was well connected to or well respected by the community of other migrants, he turned out to be a "good" seed and managed to initiate a long recruitment chain.

Number of seeds

Currently, there is no precise method for choosing the "correct" number of seeds. Past RDS studies have used from one to twenty seeds or even more (Johnston, 2008; Malekinejad et al., 2008; Montealegre et al., 2013). A small number of seeds relative to the calculated sample size may promote the development of long recruitment chains, which may produce a more representative sample. With a large number of seeds, the sample may end up with wide and short waves, resulting in a mix of recruits with characteristics similar to those of the seeds. For example, if the sample size is 500, it could be reached much quicker and with shorter recruitment chains if we start the recruitment with fifteen seeds as opposed to three.

However, heavy reliance on only one or two seeds may result in respondents with certain characteristics being under- or non-represented in the sample composition. If there are severe bottlenecks and clustering in the target population, the characteristics of the seeds may influence the composition of the final sample. For example, a survey conducted in Ukraine in 2011 among Men having Sex with Men (MSM), started with only one initial respondent, as it proved very difficult to find seeds since there were no services catering to this highly stigmatized and hidden population. The researchers intended to find the second seed in the course of data collection; however a suitable respondent was not found and the entire sample was produced from one seed who eventually recruited up to 14 waves of respondents. At the end of the survey, the sample was composed only of young, low-income individuals – resembling the characteristics of the seed. The group of higher socio-economic status, the middle-aged and older individuals, were not included in the sample. There was an obvious bottleneck around the income and age of the target population, which, in turn, was not the sample intended. Therefore, the survey data represented the network of only young, low-income MSM, rather than the entire population.

This does not mean, however, that a survey cannot start with only one seed and still provide some representation of the entire target group being sampled. In the *THEMIS-Portugal* survey, 207 Moroccan migrants were recruited from only one seed, who had extended contacts in the community: among both the recent and earlier arrivals, men and women, documented and undocumented. The recruitment chains were long enough to ensure that there was no correlation of characteristics to the seed with the outcome of recruitment (in contrast to the study of the Ukrainian survey described above), and the recruitment process penetrated deeply into the population of Moroccan migrants in the region. For instance, in Figure 4.2, the recruitment chain on the left, males (grey) and females (black) are not as well interspersed as those in the recruitment chain on the right. Nevertheless, it is generally recommended to use more than one seed for RDS surveys.

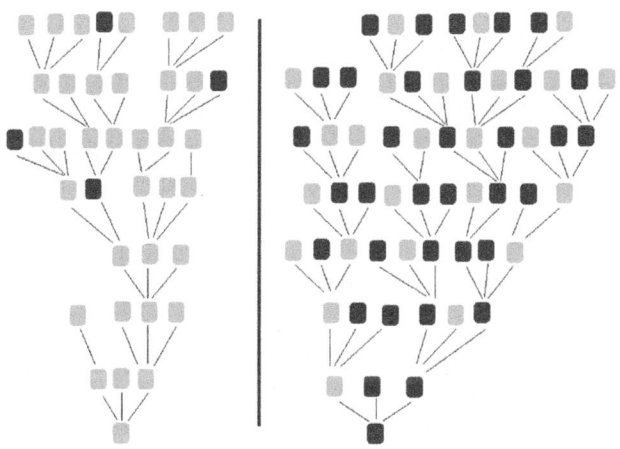

FIGURE 4.2 *Recruitment of men and women with different outcomes of samples achieved from one seed*

Legend: grey – male, black – female.

Source: Authors' simulation.

The planned sample size is therefore an important factor in influencing the number of seeds. A common-sense approach is to select the number of seeds relative to the calculated sample size, allowing for possible long recruitment chains, and to ensure equilibrium (see Chapter 7). For example, in the *THEMIS-UK* survey of Ukrainians (n = 200), three seeds were selected in the beginning and one more was added in the middle of the survey to make up for a chain where recruitment had stopped.

The number of seeds, therefore, depends on the calculated sample size, the population structure (bottlenecks), and, more practically, on the resources available and the timeframe for the survey. It is also important to bear in mind that some seeds may, in RDS language, *die out* – not recruit anyone or produce only short recruitment chains. In that case, selecting additional seeds may be necessary (see Appendix 1). If the research topic is delicate, respondents may not recruit one another, and this can necessitate the introduction of additional seeds. For example, in the *Polonia in Reykjavik* survey (n = 480), only two out of four seeds recruited others, resulting in two seeds being added in the course of the survey.

How seeds work – script for recruitment

The seeds start the initial recruitment of respondents beyond the direct control of the researcher. It is therefore important that the seeds are knowledgeable about the significance and quality of the research, as they constitute the researchers' social capital in the target community. In the *THEMIS-UK* survey of Brazilian, Ukrainian and Moroccan migrants, the researchers met with the seeds on several occasions ahead of the survey to discuss the research, and for training sessions. The seeds and the researchers knew and trusted each other and were consulted on many aspects of the research.

In order to recruit respondents, seeds need to gauge others' interest by explaining the research in an interesting and concise manner. Communication skills are therefore an important characteristic for the seed. If the seed is interested in the research itself, it will be easier for them to explain the survey aims, objectives and conditions of participation to potential respondents. On the flipside – if a seed is enthusiastic about the survey, but cannot clearly communicate this to others, they might not be the best choice for a seed.

These practicalities are best addressed during a special training session(s), where the seeds familiarize themselves, with the help of the researchers, with a script for recruitment. There, the seeds are presented with a number of coupons and asked to distribute these randomly within their network. They are asked to pass the coupon to a fixed number of the colleagues, friends, and acquaintances who are part of their social network, and who fulfill the survey eligibility criteria. The selection criteria – for example: immigration status, time of arrival, residence within a specific geographical area (a city, metropolitan area) – should be convergent with the specific network parameters included in the question on personal network size (see Chapter 3). It is important that the script for recruitment is delivered to the seeds in easy-to-understand language, as it may be used as a "template" for recruitment of all subsequent survey respondents. For example, in the survey among *sub-Saharan Africans in Morocco*, the researchers relied on the following script:

> Here are three (two or one) coupons for you to use to recruit other people you know and who know you, who are also from francophone sub-Saharan African countries living in an irregular administrative situation in Morocco. Let's go back to the question about how many people you know and they know you from francophone sub-Saharan African countries living in an irregular administrative situation in Morocco that you have seen in the past one month. The number you gave was xx. Can you think of three (two or one) people you thought of in the question above to whom you can give your coupons? If possible, try and give the coupons to different types of people who you know (e.g., different ages, different levels of income, from different sub-Saharan countries, males and females, etc.). When deciding to whom to give the coupons, please do not give any coupons to strangers.

The script above covers the following important points: the types of individual(s) to recruit, that recruitment should be from the respondents' personal network, that the respondents should start to think about who they want to recruit before leaving the survey site, and that recruitment should proceed among those with diverse characteristics.

Conclusion

This chapter has examined the selection and role of the seeds in RDS surveys. It demonstrates that, in order to strategically select the seeds, researchers should know their target population in order to bridge potential bottlenecks and to achieve a sample composed of a diverse mix of recruits. The chapter also discussed the role of seeds in the day-to-day sampling process, their possible recruitment patterns, and how potential biases could be introduced to the sample. We have essentially argued that the strategic selection of the "correct" number of communicative and respected seeds proves crucial for the successful completion of an RDS survey. These individuals ideally have large social networks made up of diverse people, can reach across bottlenecks and recruit others who will participate in the survey. Given the social and historical fragmentation within many migrant populations, bottlenecks may be present; however, these can be addressed through the strategic selection of seeds from different backgrounds.

5
Deciding on and Distributing Incentives in RDS

Guri Tyldum, Leila Rodriguesz, Ingunn Bjørkhaug and Anna Wojtynska

Abstract: *The double incentive structure is a central component of RDS methodology. This chapter explains why incentives are important, and presents issues that should be considered before deciding on what incentives to use. We address how incentives can be made culturally acceptable for the study population, the dangers of having incentives that are too high or too low, and how the distribution of incentives can be organized in a way that addresses both the security of participants and the formal requirements of project accounting. The final section addresses the ethical issues that are raised with the use of incentives in the recruitment of respondents.*

Tyldum, Guri and Lisa G. Johnston, eds. *Applying Respondent Driven Sampling to Migrant Populations: Lessons from the Field*. Basingstoke: Palgrave Macmillan, 2014.
DOI: 10.1057/9781137363619.0012.

Introduction

Respondent Driven Sampling (RDS) relies on a double incentive structure, where survey respondents receive one incentive for participating in the survey, and additional incentives for recruiting new respondents. Economic incentives are increasingly used in ordinary surveys to boost participation, but in RDS such incentives have additional functions, and are in many ways integral to the survey methodology. In this chapter, we consider the following questions: What is the purpose of incentives in RDS? What type of incentives should be used, and how can the incentives be contextualized to the study population? How do we organize the distribution of incentives in practice? And what are the ethical considerations associated with the use of incentives to recruit respondents to our study? Decisions about what kind of incentive to use depends on the characteristics of the particular project, population, and research site. This chapter addresses the different types and levels of incentives and some means to assess their appropriateness.

Motivating survey respondents to take part

Participation rates in surveys tend to vary with individual characteristics. Studies on participation rates in social surveys in general, indicate that an individual's human and social capital (e.g., educational level and community involvement) tend to boost participation rates (Tyldum, 2012; Groves et al., 2000). Participation also varies with the characteristics of the survey itself; the more interesting and relevant individuals find the survey, and the fewer barriers there are to take part, the higher the chance that they will agree to be interviewed. Participation rates also increase if economic or material incentives are offered. (Groves et al., 2004; Heberlein & Baumgartner, 1978; Roose et al., 2007).

RDS is based on the premise that peers are more effective than outreach workers and researchers in locating and recruiting members of a hidden population (Heckathorn, 1997). In classic survey sampling methods respondents are typically approached in the place where they live or work, and only need to take the time to answer a set of questions. In RDS surveys on the other hand, potential respondents, who have received a coupon that makes them eligible to take part, need to initiate contact with the survey organization, travel to the interview site, and recruit other respondents. This means that an RDS survey asks more of its respondents than traditional sampling methods. A successful RDS survey therefore needs respondents that are motivated to take part and recruit, and the double incentive system is a vital part of this. Each respondent is given an incentive to participate (primary incentive), and another for the eligible peers they recruit and who enroll (secondary incentive).

Primary incentive

The primary incentive works as a remuneration to show respect and appreciation for an individual's time and effort in participating in the study (Semaan et al., 2008). There are myriad reasons why people may prefer not to participate in a survey. For many it is simply a matter of priorities in a busy daily schedule, and the aim of an economic incentive is to motivate respondents to set aside the time necessary to take part. Although, many respondent groups might participate in surveys without any incentives being offered, RDS surveys are often conducted in marginalized populations (who may not often have a chance to voice their opinions), many of whom may appreciate having the opportunity to take part in research. This means that parts of the population may be willing to participate without economic incentives. However, in order to produce representative data, it is necessary to reach all population groups, and not only the most motivated. And as RDS is usually conducted in populations where no sample frame exists, we have few ways of accessing the extent to which we reach all sub-groups of a population. The incentive can, however, reduce potential bias created by volunteerism. It can also facilitate participation for people who have economic barriers to participation, to compensate for income losses or transportation costs.

Secondary incentive

Douglas Heckathorn (1997) claims that the secondary incentive is the most important in securing recruitment in an RDS survey. Being recruited by a friend or acquaintance can create peer pressure or altruism that could motivate participation among people who otherwise would not have taken part in a survey. This means that people do not only participate for their own economic benefit, but also because they know that somebody else will benefit from it. This was illustrated in the 2006 *Polonia in Oslo* survey (see Appendix I, for a thorough presentation of all surveys referred to in this volume). A relatively wealthy business leader came in to be interviewed, but declined the primary incentive, arguing that he only came in because he wanted to do one of his employees, who recruited him, a favor. After the interview, he accepted the coupons, and a few days later, he also recruited someone, probably also seeing this as a favor to that person, as the new recruit then got an opportunity to earn some extra money. This way, the primary and secondary incentives work to motivate the participation of different population groups.

Determining the type and value of the incentive

Determining an adequate incentive (type and amount) is central to the successful implementation of any RDS survey. The incentive should be high enough to motivate all sub-groups of the survey population to participate. But if the incentive is too high, there is a risk of motivating non-eligible persons to pretend to be part of the population (masquerading, see Chapter 6) in order to receive the incentive, or even selling or bartering coupons outside of networks.

Incentives are usually determined through the initial formative assessment (see Chapter 6). The incentives should be culturally appropriate to the study population. The optimal incentive will depend on both the cost of living and the average income for the population, as well as on the efforts expected from the respondents to participate in the survey, including the time and cost of travel to the survey site. Ethical guidelines and even a country's legislation might further shape the incentives selected.

Compensating for time use in line with average salaries for the group

Several surveys have used a primary incentive approximating the average hourly salary for the population. This was the case in the *THEMIS-Norway* survey, giving NOK 150 (about 19 Euros) as the primary incentive for an hour-long interview, and NOK 100 (about 13 Euros) per recruited respondent as the secondary incentive. The *THEMIS* project was conducted in parallel in several countries, but as salaries are higher in Norway compared with the other countries involved, the other survey organizations chose lower incentives, adapted to context-specific conditions and local living costs.

In an RDS survey among ex-combatants in Liberia, many former combatants worked on day-to-day contracts on small farms, and in order to participate in the survey, they would lose a whole day's work in the fields. The survey organization offered combined primary and secondary incentives almost equivalent to what could be earned from one day's work in the field (1.78 Euros). If they had not been compensated for a day's pay, many respondents would not have been able to afford the time off work. However, a day's salary was attractive, and even those not eligible for the survey tried to enroll. Anticipating this, the survey staff had recruited local NGO staff that were themselves ex-combatants, to screen for eligibility, asking a number of control questions before enrolling respondents. If the recruited peers were ineligible, the recruiter would not receive the secondary incentive. The nature of the survey was quickly known in the community, and after a few days, recruiters no longer tried to enroll peers who were not former combatants (Bøås & Bjørkhaug, 2010).

If there are significant income differences in the population, it is not always necessary to have incentives that approximate the incomes of the wealthier sub-groups of the study population. Income levels are often strongly correlated with educational level and social capital, and the literature on sampling bias in traditional probability surveys has shown that the effect of monetary incentives is higher for population groups with lower education and who possess low social capital (Tyldum, 2012; Stoop, 2004; Roose et al., 2007; Heberlein & Baumgartner, 1978). Groups with higher education and social capital may be more likely to participate without economic incentives because they are interested in the topic of the survey, because they feel it is a social responsibility, or for other reasons (Groves et al., 2000). The secondary incentive, which encourages recruitment through altruism and peer pressure, will also mobilize the wealthier respondents. However, if there is a strong correlation between the time available to be interviewed and a low-income level in the target population, you should monitor the recruitment closely, and consider increasing the incentive. For instance, in some migrant populations, a significant portion will be unemployed, while others work very long hours. The unemployed, having more time and being in greater need of income, will be relatively easy to recruit for an RDS survey. The working section of the population may, on the other hand, need more motivation to take time out of their working schedule. In such a situation, it may be worth monitoring if recruiters have difficulties in recruiting employed people, by asking a few questions when they come to collect the secondary incentive (Johnston, 2013b). If such biases seem to appear, increasing the incentive may be an appropriate response.

Stratified incentives

Methodologically, it is also possible to stratify incentives, and give higher value incentives to some population sub-groups than others. However, this is not recommended, as it might put the survey implementation at risk. Individuals who receive the lower incentive (for doing the same task) may feel discriminated against and discouraged from participating or may pretend to be part of the population that gets the higher incentive (Johnston, 2008; 2013b).

Differentiating on the secondary incentive is less problematic in terms of impact on the survey organization, but should also be avoided, as there is a risk of oversampling the groups that receive extra. If the distribution of the characteristics in the population is unknown, it can be difficult, or even impossible, to monitor what is sufficient or excessive encouragement. That said, if significant bottlenecks are discovered in the recruitment process, stratifying the secondary incentives can be considered to motivate respondents to recruit across the bottleneck. For instance, in some populations, men and women may not recruit across. This is not because men and women do not know people of the opposite sex, but rather that they tend to recruit those they interact with socially, and this tends to be stratified by gender. In such situations, it could be worthwhile to pay a higher secondary incentive to motivate people to recruit persons of the opposite sex. If so, respondents should be asked about the gender composition of their personal social networks, to allow you to assess the effect of the stratified incentive. If the portion that recruits persons of the opposite sex is higher than the share of opposite sex individuals in their social network, the level of encouragement is too high.

The impact of incentives that are too high or too low

Enrolment biases such as masquerading, repeat participation and the selling and bartering of coupons are the most common problems if incentives are too high (in addition to the ethical issues discussed later in the chapter). Selling and bartering coupons breaks with both the important first assumption that the population form reciprocal relationships and the fourth assumption, that recruitment takes place at random within the respondents' networks (see Chapter 2). Some survey organizations choose to introduce eligibility checks in response to attempts at masquerading (see Chapter 6).

Many surveys include a question to monitor the relationship to the person's recruiter, and if several respondents report being recruited by a stranger, it may be an indication that incentives are too high (Johnston, 2013b). Lowering incentives will reduce the need to rely on screening and reduce the risk of coupons being sold to strangers. It may also be appropriate to have a lower primary incentive and a higher secondary incentive, as they do not have the same direct effect on enrollment biases, particularly in studies where respondents are expected to have large social networks of eligible people from which to choose.

Different problems occur when incentives are too low. For instance, in the 2006 *Polonia in Oslo* survey, after weeks with no response from the initial three seeds, the primary incentive was increased from NOK (Norwegian Kroners) 100 to 150 (about 13 to 19 Euros), and the secondary incentive from NOK 50 to 100 (about 6 to 13 Euros). At the time of the study, two-thirds of the Poles in Oslo worked in the construction sector, where the average hourly wage for the total workforce was then NOK 174 (about 21 Euros) (Friberg & Tyldum, 2007). However, most of the Polish workers earned significantly less than this, and NOK 100 was closer to the average net salary for this group. At the same time, more than half of the Polish construction workers worked over 50 hours per week (Friberg & Tyldum, 2007), and many valued their spare time more than the survey incentive. Higher incentives were therefore needed to encourage these workers to take time to travel downtown for an interview after a ten-hour working day. In parallel with the incentive increase, the coupon and recruitment procedures were simplified. The result of these two changes was that the recruitment quickly picked up.

Although it may be necessary to either increase or decrease an incentive after the survey begins, efforts should be made to anticipate an adequate incentive that will be employed throughout the entire data collection period. If the incentive must be changed, it is best to do so as early as possible.

In a case where the incentives do compensate for transport costs and respondents' time, in line with average hourly salaries for the group, but recruitment still does not work, it is not necessarily the value of the incentives that is the problem. Rather than rushing to change the incentives, it may be worth taking a closer look at the fieldwork organization to see if any changes can make participation easier, or otherwise, to improve the experience people have when they come in for an interview (see Chapter 6).

Non-monetary incentives

Some researchers are uncomfortable about handing out money for survey participation, and feel that giving food, lottery tickets, telephone cards or other concrete/material gifts carries more positive connotations and are easier to defend ethically. Giving out gifts can also reduce security risks and be easier for administrative and accounting reasons. However, using cash incentives has the advantage of having a broader reach; everybody can use money, while not all may be equally interested in, for instance, cinema tickets, phone cards, or a particular type of food.

RDS without material incentives

Some RDS surveys have not used any material incentives, either because of budgetary constraints or out of ethical or practical considerations. In RDS surveys, incentives help to reduce the bias introduced by volunteerism (e.g., individuals who have a lot time at their disposal or who have a particular interest in the survey topic). In studies where material incentives are not used, it is particularly important to monitor recruitment patterns, in order to assess the extent to which there are systematic biases in the data produced.

In the absence of material incentives, it is necessary to ensure that as much as possible is done to induce recruitment drawing on other factors. In the *Nigerians in NYC* survey, the ethical review board discouraged the use of economic incentives. The survey organization was therefore altered to minimize the burden for the respondents, with flexible fieldwork sites, where the interviewer travelled to an area convenient for the respondents to meet instead of them coming to a survey site. As Nigerians in New York City are often active members of a church or mosque, or of hometown or ethnic associations, the researcher contacted the potential gatekeepers of the community – pastors, imams and association leaders – and asked them to endorse the research. In this situation, the strong endorsement by community leaders may have "incentivized" the population to participate.

There have also been several unsuccessful attempts to conduct an RDS survey without material incentives. For instance, in the survey among ex-combatants in Liberia, the local implementing partners and NGOs were initially skeptical about using monetary incentives, and wanted to try recruitment without any incentive first. However, after the initial three seeds had been interviewed, none of them recruited any peers to the survey. Cash incentives were then introduced, and recruitment quickly picked up (Bøås & Bjørkhaug, 2010).

Making participation a positive experience

A positive interview experience can also work as an incentive to recruit new respondents. If respondents experience the survey as meaningful and important, feeling secure and appreciated when they come in for the interview, they are likely to convey this to their friends. However, if their experience is frustrating or downright negative, they may be reluctant to send their peers, and hence successful recruitment will be much harder to achieve.

Much can be done to make survey sites welcoming, to give respondents a positive experience. In the *Polonia in Oslo* surveys, respondents were offered as much coffee, tea or hot chocolate as they wanted, and could sit down to read Polish newspapers and magazines before and after interviews. As substantial parts of the population lived in barracks or as au pairs in their employers' households, many appreciated this opportunity, and on the weekends, the waiting room in the survey site virtually turned into a Polish café. Staff training should emphasize the importance of treating respondents (who often belong to marginalized groups) with respect.

Organizing the distribution of incentives

A major advantage of RDS surveys is that they can secure fully anonymous participation for respondents. Because respondents in an RDS survey initiate contact themselves, we have no other information about them, beyond what we know about their recruiter and what they themselves provide. This does, however, raise problems for some accounting departments, as they may want to have contact information and signatures in order to allow the hand out of money or gifts. There are, however, several different strategies for incentive distribution that accounting departments may accept. Several surveys have collected the coupons (with unique numbers), as a receipt, and used these in combination with data files to document that the money had actually been distributed (Friberg & Tyldum, 2007; Johnston, 2013b). This method requires that the coupon has two sections, one that can be handed in for the primary incentive, and one for the recruiter to keep for the collection of the secondary incentive (see, for example, the coupon from the *THEMIS-Norway* survey in Figure 6.3, Chapter 6). Other survey organizations had interviewers or supervisors sign a log, either the coupons or a separate document each time they distributed incentives. Even so, some accounting departments prefer that researchers distribute gift cards or material gifts instead of cash. The *THEMIS-Norway* survey used "Universal Gift Vouchers", which can be used in most stores and businesses throughout Norway, in order to avoid registering the payments (and to whom the payments were made) with the tax office.

There may be cultural and class differences in how individuals view receiving money. While some individuals do not mind receiving cash up front, others will appreciate receiving it more discreetly. Money can be put in an envelope that is handed to the enrolled respondent before the interviews start, as this might give a stronger connotation of handing a gift in appreciation of them coming in. Having money pre-packed in envelopes can also reduce security concerns of having visible cash. Staff should also be trained to explain to respondents the reason for the incentives, and to express their gratitude for them taking time to come in for an interview. Staff should be trained to recognize that even though they give respondents money for participating, efforts are still needed to make respondents feel respected and comfortable in the interview situation.

To collect the secondary incentive, recruiters usually have to revisit the survey site. Some surveys use the opportunity of the second visit to conduct a follow-up interview with recruiters, in order to map any challenges in the recruitment process. This interview can be done qualitatively, as part of parallel monitoring, or more systematically to give analysis of non-response (Johnston, 2013b). The collection of secondary incentives can however, create logistic challenges for the recruiter, for instance, in populations where many individuals work long hours, or live far away. In the *Polonia in Oslo* survey, it was possible for recruiters to ask their recruits to collect the secondary incentive for them by sending the top part of the coupon along with their recruit. This saved respondents time from making a second visit to the survey site. Although it restricted the opportunity to do follow-up interviews, it made the recruitment work smoother, with less effort involved for recruiters motivated by the incentive.

The ethics of incentives

Although economic incentives have been widely used in recruitment of respondents for medical and psychological research, it is still relatively new in the social sciences. Many researchers feel ethically concerned with employing incentives as a recruitment tool. If individuals do not want to participate in a survey without an incentive, is it then ethically defendable to persuade them by offering payment? The answer is not straightforward, and in this section, we discuss some of the main dilemmas to consider before deciding if RDS can or should be used for recruitment. If your institution or country has ethical guidelines, you should also familiarize yourselves with those.

Offering an economic incentive can be considered as taking a step away from the ideal of free and informed consent in survey research (Tyldum, 2012; Grant, 2006). Some choose to offer respondents the primary incentive up front, and not upon completion of the questionnaire, to avoid that payment is understood as conditional on answering the questions, or that the respondents fears s/he needs to answer the questions "correctly" in order to receive the incentives. Incentives should not induce individuals to act in ways they would not consider under other circumstances, or to disclose information about themselves that they wish to keep private. Essentially, an economic incentive should not be so high that some sub-groups of the target population feel they cannot afford not to take part. The aim of the incentive is to make participation more attractive to people who would rather spend their time otherwise, but who do not mind being interviewed. Anyone who does not want to participate in a survey, however, whatever their reason, should feel able to refuse. As a general rule, an economic incentive should not be more than a payment for the equivalent time spent working for the group's average hourly salary (Grant, 2006; Macklin, 1981). Importantly, respondents need to be informed that payment is not conditional on answering all questions. And, as in any regular survey, informed consent should be obtained before the interview starts.

A survey using incentives to increase participation needs a stronger awareness of risk of harm than other surveys (Tyldum, 2012). Particular care should be taken when recruiting people from groups who are linked to illegal or strongly stigmatized behavior (Brunovskis & Bjerkan, 2008). Even if the researchers themselves feel that they have done everything they can to secure confidentiality, respondents may still feel insecure. There is also no guarantee that respondents' membership in a stigmatized target population will not be recognized if they are seen entering the survey site or carrying a coupon.

The risk of harm not only pertains to the risk of being identified as part of a stigmatized population, but also to the specific survey questions. Questions that entail a risk of re-traumatizing respondents, or that may evoke painful memories, should be avoided if possible. In some surveys of traumatized and stigmatized groups, such as people who inject drugs, sex workers, men who have sex with men, and youth, sensitive questions are asked (Johnston, 2013b). In these cases, survey coordinators often plan to have a trained professional at the RDS interview site, who can respond to respondents' concerns or refer them to adequate services if follow-up care is needed. However, if professional follow-up of respondents who might need this cannot be secured, questions that involve a risk of re-traumatizing respondents should be avoided.

Conclusion

Clearly, incentives are integral to the recruitment of respondents in an RDS study. Monetary incentives are most commonly used, but some surveys rely on alternative incentives such as gift cards or other small relevant items of value to the study population.

Different populations pose distinct challenges in determining the best incentive to use. Wealthier or more highly-educated immigrant groups may not be motivated by the modest financial incentives commonly used in RDS, but can respond well to alternative incentives, peer pressure, a wish to be heard, or a desire to contribute to research. Respondents with lower incomes may respond well to modest financial incentives, but there is no guarantee that the planned incentive will work well. Signs that incentives are too high include, masquerading (attempts to enroll by individuals who are not members of the target population), and the selling and bartering of coupons. Incentives that are too low can result in slow or no recruitment. High incentives can be lowered if resentment in the population of interest can be avoided. Low incentives can be increased, if budget permits, or they can be complemented with alternative, non-material incentives. It is important, however, to get the incentive right from the beginning, or change it as early as possible if necessary, as changing incentives also affects the probabilities of recruitment.

Taking into account ethical considerations in the use of incentives, as well as budgetary constraints, prior knowledge of the population of interest, and some creative thinking, incentives are not only a central aspect of RDS methodology, but a reward to respondents for making research possible.

6
Formative Assessment, Data Collection and Parallel Monitoring for RDS Fieldwork

Jane Montealegre, Antje Röder and Rojan Ezzati

Abstract: *This chapter addresses the key aspects of RDS surveys that need to be planned in advance of fieldwork (including coupon design, survey site location, and project staffing), as well as possible strategies for monitoring recruitment once the survey has started. We show that such parallel monitoring is crucial in order to learn how recruitment plays out in the population and to determine whether any adaptations are needed. This chapter also provides an overview of the type of adaptations that are commonly made to tailor the survey to the target population (including strategies to address slow recruitment). Using examples of RDS surveys conducted among migrant populations, we show that survey logistics can, and often need to, be adapted to ensure the success of RDS recruitment.*

Tyldum, Guri and Lisa G. Johnston, eds. *Applying Respondent Driven Sampling to Migrant Populations: Lessons from the Field.* Basingstoke: Palgrave Macmillan, 2014. DOI: 10.1057/9781137363619.0013.

Introduction

As in all survey research, RDS data collection is the culmination of months of careful planning. Nonetheless, it can be difficult to predict how an RDS survey will unfold, especially when the method has not been used before in a particular population. We generally do not know in advance to what extent our target population will respond to the sampling method, or to what extent the survey characteristics, such as the incentives and the hours of operation, will suit the needs of our respondents. Thus, getting RDS to work often involves an iterative process of observing how well the method works and being prepared to modify certain elements to better tailor it to the population. We call this parallel monitoring. In this chapter, we discuss key aspects of an RDS survey that need to be planned in advance of data collection and show strategies for monitoring the recruitment once the survey has started. Common questions may be: How do I choose the survey location? How do I decide the appropriate number and type of staff? Should I use a fixed- or mobile-site? We then provide an overview of the type of adaptations that are commonly made to tailor the survey to the target population. We begin with narratives of two RDS surveys to better illustrate the type of challenges researchers face and the type of solutions that can be adopted. The first is the *Central American Women in Houston* survey (see Appendix I, for a thorough presentation of all surveys referred to in this volume). The second is from the *THEMIS* surveys. Both case studies portray how RDS requires careful planning and ad hoc flexibility in adapting the method to different populations and circumstances. We draw on these and other examples throughout the detailed discussion that follows.

Case Study #1: Survey among Central American Women in Houston

> This survey sought to describe the prevalence and context of HIV risk behaviors among undocumented Central American Women in Houston. The social networks that are so critical for immigration to the United States gave us a sense that RDS would be a good method to survey this group. On the other hand, we were concerned that the personal networks among recently arrived immigrants may not have been sufficiently dense to sustain RDS recruitment, and that bottlenecks among immigrants of different national or ethnic origins could preclude recruitment across groups. Furthermore, we

worried that fear of deportation and discrimination could hinder participation. Thus, as often occurs with RDS, we embarked on an experiment to evaluate whether RDS would work and under what conditions.

We began with a formative assessment to learn about the social network characteristics of Central American women (e.g., the personal network size and frequency of social encounters) and to define the practical aspects of the survey. This assessment involved multiple activities, including a review of the immigration literature, field observations, in-depth interviews and ethnographic mapping. Based on this information, we decided to use a classic RDS fixed-site approach and set up a data collection space within the office of the local immigrant service organization (ISO), and that our hours of operation would be from 2 to 9 p.m. to accommodate working women. We used cash incentives, and the coupons specified that recruits should call to schedule an appointment.

On our first day of fieldwork in February 2010, we met with two women who were recruited by the ISO staff to serve as our seeds. We explained the recruitment process, motivated them to join the project, gave them their coupons, and sent them off to start recruiting. After months of preparation, we had to let go of some of our control and trust in our seeds to initiate the recruitment process. During the first week, we had two recruits. During weeks two and three, recruitment remained stagnant. During week four, we decided that we needed to plant a new seed and recruited a gregarious Salvadoran woman, with hopes that she would reignite recruitment. Nevertheless, recruitment continued to proceed slowly (with an average of 6.3 recruits per week), and by week six, recruitment came to a halt.

After four days without a single interview, we concluded that we had to take action. Fortunately, in the previous four weeks we conducted parallel monitoring, which involved talking to respondents and other informants and walking around the neighborhood to observe people's interaction in order to try to understand why recruitment was not proceeding as planned. In doing so, we learned several things. First, although we considered the survey site at the ISO to be centrally located in the immigrant neighborhood and accessible via public transport, we found out that some women were hesitant to enter the apartment complex to reach the site as this involved passing a security guard at the entrance gate. Second, although in-depth interviews conducted prior to recruitment indicated that respondents preferred nighttime hours of operation, we discovered that many women feared leaving their homes at night due to criminal activity in the area. Third, we noticed a general lack of understanding of the recruitment process. Fourth, women seemed hesitant to call to make an appointment, and those who did often missed their appointment.

In response, we made several changes. We established a new interview site at a street-side apartment with the hours of operation from 9 a.m. to 5 p.m. The original site remained open with shorter hours of operation. We began to explain recruitment more slowly and use diagrams to explain the recruitment process, and we no longer asked respondents to call to make appointments. Within a week of making these modifications, the rate of recruitment increased to 29 respondents per week, a pace that was largely sustained until near the end of the survey. By the end 16-weeks, we recruited 226 non-seed respondents, exceeding our target sample size of 180. We attribute our success to our parallel monitoring activities, which allowed us to determine the causes of the faltering recruitment, and allowed us to adapt.

Case Study #2: The THEMIS surveys

Additional layers of complexity are introduced when parallel studies are implemented with multiple target populations in multiple geographic areas. This was the case for the *THEMIS* survey of Brazilian, Moroccan, and Ukrainian migrants in four European countries: Norway, Portugal, the Netherlands, and the United Kingdom. The project studied the link between these groups' migration patterns and different types of migrant networks in both the country of origin and the country of settlement. RDS was chosen as the sampling methodology as it relies on the populations' social networks and would, in itself, provide valuable data on the size and types of networks of these populations.

While the survey involved three different populations in four countries, it is important to note that a separate RDS survey (with its own system of coupons and set of seeds) was implemented for each of the target populations. Here, we focus on the Brazilian part of the *THEMIS-Norway* surveys. We started the survey using a fixed-site approach. Respondents were invited for an interview at our research venue outside of regular office hours (on Saturdays and two afternoons a week). However, five weeks into the survey, we realized that recruitment was too slow. As a result, we decided to change to a mobile-site and to have respondents call to schedule an interview at their preferred time and location. The only restriction was that the interviews were to be conducted in public areas, such as at our offices, cafés or restaurants, due to safety precautions for the interviewers. Staff for the mobile-site consisted of a "Telephone Manager" who scheduled interviews; interviewers, who met with respondents at the chosen time and location; and a "Front Desk Manager" who, during specific opening hours at our offices, distributed secondary incentives. Our respondents expressed their appreciation of the flexibility provided by the mobile-site and of being able to adjust the times of the interviews to their individual schedules.

Other changes we made included adding seeds and allowing respondents to send coupon numbers electronically by SMS or e-mail to their recruits. We also moved from two RDS coupons to three. These changes were implemented as the result of parallel monitoring during data collection.

Planning and formative assessment

Formative assessment will vary from survey to survey, depending on cost and time constraints, as well as the researchers' degree of familiarity with the population. In some studies, it is sufficient to conduct a rapid formative assessment. For example, in the *sub-Saharan Africans in Morocco* survey, the formative assessment was conducted over the course of three days. During this time, the research team conducted interviews with migrants and organizations that serve migrants to inform them of specific decisions regarding the survey design. At the opposite end of the spectrum, other studies conduct extensive formative assessments. For example, in the *THEMIS-Norway* survey, researchers became familiar with the target populations in Oslo through previous in-depth qualitative studies conducted during the two years prior to implementing RDS. While it was not the purpose of the qualitative studies to inform the RDS survey per se, the first-hand knowledge gained was essential in guiding the design and field logistics of the RDS surveys.

The formative assessment activities of most RDS surveys lie somewhere in between the two examples given above. In the *Central American Women in Houston* survey, the formative assessment phase involved three key activities: ethnographic mapping to learn about the spatial patterns of Central American settlement in Houston; interviews with members of the target population to learn about pertinent social network characteristics (e.g., personal network size and potential barriers to social mixing across sub-groups); and focus groups with key informants and members of the target population to determine a site location and opening hours. Formative assessment then carries on into *parallel monitoring*. As we will discuss later in this chapter, ongoing parallel monitoring, which is conducted during data collection and involves similar methods as formative assessment, is important for continually learning more about the target population and adapting the survey logistics.

However, before starting data collection you need to make a number of decisions including choosing a survey site, staffing, designing coupons and deciding when to start the survey. We discuss these below.

Survey sites

RDS surveys can collect data in a fixed- or mobile-site, or through the internet (Johnston, 2013b). The fixed-site approach means that respondents visit a specific location to complete the survey and to collect their secondary incentives. Data collection may be done on a walk-in basis or using scheduled appointments. The mobile-site approach means that interviewers meet respondents in changing locations, often selected to suit the respondent. Finally, the web-based approach does not involve the use of a physical site since data collection and disbursement of incentives are done entirely online (Wejnert & Heckathorn, 2008). We limit our discussion below to fixed- and mobile-site approaches, given that the application of web-based RDS is still in its infancy.

Fixed-site approach

When using a fixed-site approach, an important consideration is the location of the survey site. Locations such as a commercial storefront apartment or an existing office space of an organization that serves the target population are often used. In the *Central American Women in*

Houston survey, the main survey site was an apartment near a commercial area in a Central American neighborhood. In the *Polonia in Oslo* surveys, the site was located in the research institution's offices in an immigrant neighborhood. Either way, it is crucial that the survey site be easily accessible to members of the target population, ideally within close proximity to where they live or work, and/or accessible using public transport. The neighborhood should also be reasonably safe for both survey staff and respondents (Johnston, 2008; 2013b).

It is also possible to have multiple sites, and a decision to do so is based primarily on the size of the target geographical area and the potential barriers to mobility. It may be advantageous to have multiple sites if the sampling area is large or if public transportation is costly or time-consuming for respondents. On the other hand, multiple sites can introduce additional complexities, for example, with coupon numbering, tracking who recruited whom, and distributing secondary incentives. It is necessary for recruitment to cross-over between multiple sites, in order to ensure that the sample comprises a single social network component (Johnston, 2013b).

When multiple sites are used, it is not necessary to operate all sites at the same time, and some studies designate a rotating schedule. For example, in one RDS survey on HIV and high-risk behavior, the research team worked at the main study office on Mondays, Wednesdays, and Thursdays, at a community center located in a public housing project on Tuesdays, and at another community center on Fridays. Multiple sites were deemed necessary given the geographic dispersion of the target population in a large city and the limited availability of public transport. Furthermore, by rotating the same survey staff across multiple sites, the project was able to make participation convenient without introducing additional staffing costs (Risser & Montealegre, 2013). However, it should be noted that rotating across multiple sites may increase the complexity of the instructions given to recruits and may not be feasible in populations with very low literacy. Opening hours and multiple site addresses should be clearly marked on the coupon.

When using a fixed-site approach, it is possible to conduct interviews either on a walk-in basis or through appointments. This decision is usually based on the expected influx of respondents and knowledge of their cultural norms. For example, in the *Central American Women in Houston* survey we realized soon after commencing data collection that the women were often reluctant to make appointments, making it impossible to require recruits to schedule an interview ahead of time. In this scenario, the researchers opted to offer walk-in interviews and appointments, a common strategy in RDS studies (Johnston, 2008; 2013b).

Tailoring data collection to the target population also entails setting appropriate hours of operation. For example, when working with labor migrants, who often have long, irregular, and/or uncommon working hours, interviews at night or on weekends might be needed. In the *Polonia in Oslo* surveys, most interviews were conducted in the evenings and on the weekends to accommodate workers, particularly those in the construction and domestic work industry, who work long hours during the week.

Mobile-site approach

Some RDS surveys have used mobile sites, as exemplified by the *THEMIS-Norway* surveys described above. Another example is from the *Nigerians in NYC* survey, in which the researcher chose to arrange an interview time and location to suit her respondents because of the large geographic distribution of the population throughout the city. Interviews frequently took place in respondents' homes or workplaces and, less frequently, in public venues such as coffee shops.

Staffing

Whether using a fixed- or mobile-site approach, there are many issues regarding survey staffing that influence the success of RDS. Many RDS studies with migrants employ staff of the same ethnic and linguistic background, which can be advantageous as it generally eases rapport with survey respondents, overcomes linguistic barriers, and ensures that linguistic and cultural nuances are captured (Feskens et al., 2006). Additionally, and of particular importance in an RDS survey, eligibility screening is best done by staff who are familiar with the target population (Johnston, 2008; 2013b). That being said, finding staff who are members of the target population can sometimes be difficult. For example, in the *THEMIS-UK* survey, the researchers experienced difficulty in recruiting Moroccan interviewers. It is worth noting that in some cases respondents might also be more sensitive towards the class background or ethnicity of their co-migrants. This was the case in the *THEMIS-Norway* survey. Interviewers reported that some darker skin respondents seemed wary of being interviewed by lighter skin Brazilians, indicative of class divides in Brazil.

A second consideration is the number and roles of survey staff. This decision is usually made based on the expected intensity of recruitment, which can be difficult to foresee and may require adjustment. There are various roles for survey staff in RDS studies: respondent screening, interviewing, coupon management, crowd control and respondent flow. The allocation of staff for each of these roles varies from survey to survey. For example, in the *Central American Women in Houston* survey, only two staff members were each in charge of screening, interviewing, and managing coupons. In contrast, other surveys have designated a staff member for each of these roles, which can be particularly important if there is a high volume of respondents.

In the *THEMIS-Norway* survey, the mobile-site approach additionally involved a phone manager, who was the initial contact point for interview scheduling (Figure 6.1). The phone manager asked potential respondents questions to determine their eligibility for the survey and inquired about the preferred time, place, and language for the interview. After consulting the electronic calendar indicating the interviewers' availability, the phone manager sent a text message to the available interviewers with the time, place, and RDS coupon number. The first interviewer to agree to conduct the interview would receive the potential respondent's phone number to make further arrangements, while the other interviewers would receive a text message indicating that the interview had been scheduled.

Survey coupons

Coupons are the "face" of an RDS survey, playing the role of an advertisement flyer in addition to allowing researchers to track who recruited whom. In deciding whether to take part in the survey, potential respondents have two sources of information: what they hear from their recruiter (and potentially others in the community) and what they see on the coupon. It is therefore crucial for the coupon to give the right message. The design of the coupon is usually based on some type of formative assessment and on examples of coupons used in past surveys (Johnston, 2008; 2013b). The coupon should make the survey appear inviting, should not disclose any stigmatizing information, and should instruct potential respondents on how to proceed. Figure 6.2 shows the coupon used in the *Central American Women in Houston* survey. Note that although the target population was undocumented migrants, the

Formative Assessment, Data Collection and Parallel Monitoring 71

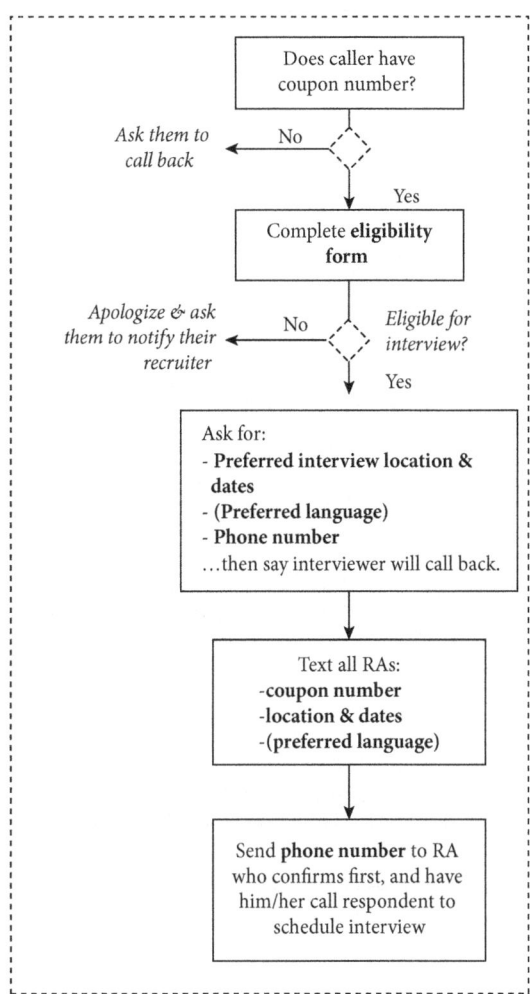

FIGURE 6.1 *Overview of respondent flow. Example*
Source: THEMIS studies.

coupon does not mention documentation status. This information was deliberately omitted given that it could potentially incriminate individuals and might discourage those who do not want to reveal their documentation status to peers. The other side of the survey coupon provides key information about the survey, including the project phone number, the address, and instructions to present the coupon at the time

FIGURE 6.2 *Survey coupon (front and back)*
Source: Survey of Central American Women in Houston (English translation).

of participation. While it may be tempting to put a lot of information about the project on the coupon, it should be kept simple and only include what is strictly necessary. Remember that the recruiter will tell the potential respondent about the survey, such as the kind of questions asked or the incentive given, so such information generally does not need to be included on the coupon.

Many RDS surveys use a two-part coupon where one portion of the coupon is given to the recruit to enroll in the survey and the other is kept by the recruiter in order to redeem the secondary incentive. In the *THEMIS-Norway* survey, which used a mobile-site approach, information regarding the location of the survey office was directed toward recruiters who were instructed to visit the survey office to redeem their coupons (see Figure 6.3).

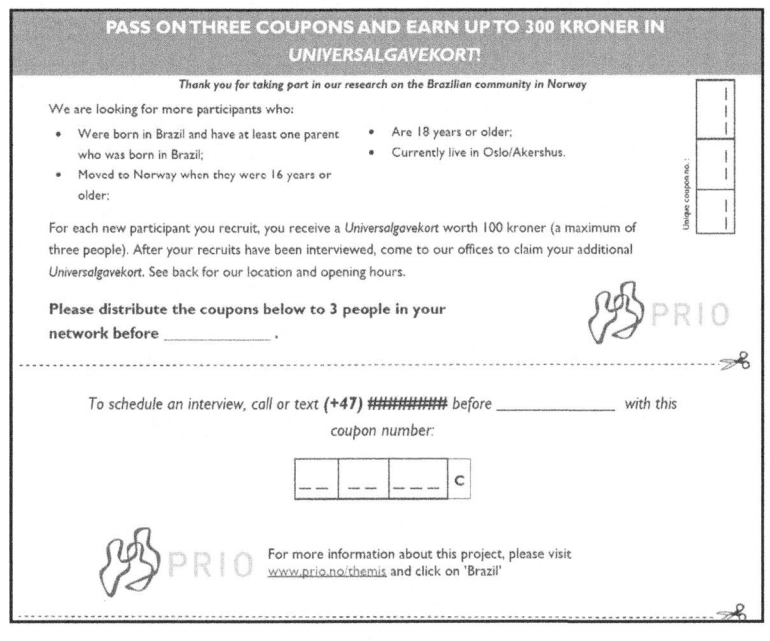

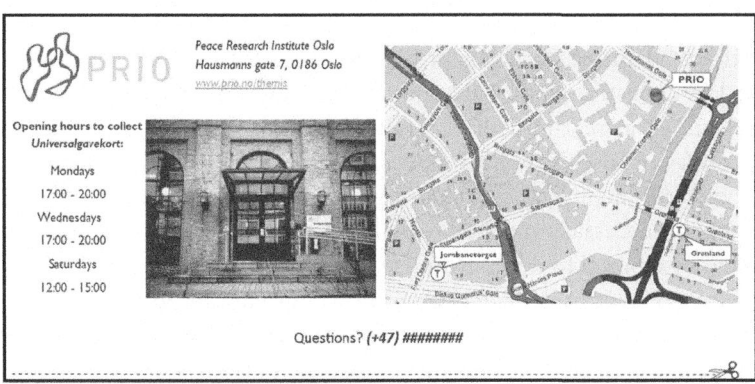

FIGURE 6.3   *Survey coupon with section to be kept by recruiter (front and back)*
Note: Notice that the front side has a portion that is kept by the recruiter.
Source: THEMIS Survey of Brazilians in Oslo survey.

Data collection start date

Although it may be tempting to start a survey as soon as the preparation is finished, selecting an appropriate start date needs careful consideration of the possible seasonal fluctuations in the composition of the migrant population (see Chapter 2). While this factor is equally present in non-RDS surveys of migrants, it is even more crucial to consider for RDS surveys, which rely on peer-to-peer recruitment. Selecting an unsuitable time to start the survey may result in a sample of only one subgroup of the population and it may be difficult to initiate and continue recruitment to achieve the necessary sample size. For example, seasonal labor migrants who work in agriculture may be present in the summer months but may leave after the fall harvest. Similarly, formative assessment for the *THEMIS* surveys indicated that the three target populations had different visiting patterns to their respective countries of origin, related to religious and summer holidays, harvests and labor patterns. These factors should affect decisions about when to start data collection to ensure reaching the target sample size before the composition of the target population changes. Be aware that recruitment may take longer than anticipated.

Data collection and parallel monitoring

Initiating data collection

After the planning decisions, it is time to start data collection. Initiating data collection is probably one of the most daunting aspects in an RDS survey, as recruitment may initially be slow. Even if recruitment is robust, it can be difficult for survey staff to deal with the unpredictability of the recruitment process. Patience at the beginning of data collection is sometimes necessary and, before making any modifications, recruitment should be allowed to develop on its own.

However, recruitment does not always develop by itself, and many studies have had to make some adjustments to make it work. For instance, in the *Central American Women in Houston* survey, numerous modifications were initiated in week 7 of data collection, resulting in a rapid increase in recruitment (see Figure 6.4). It is quite unpredictable how recruitment and data collection will unfold in RDS surveys, especially considering that the method is often used for populations

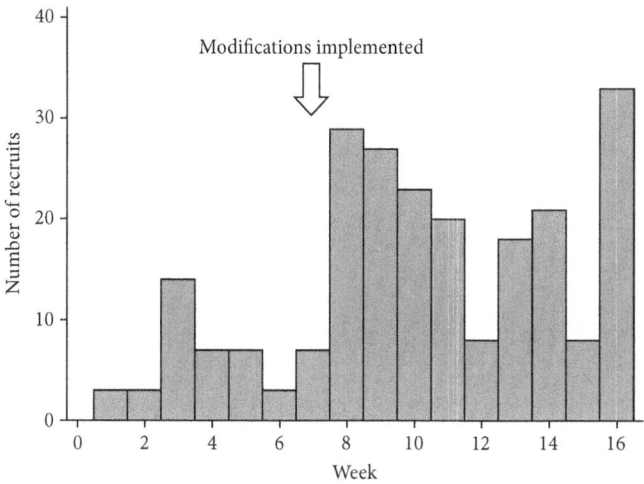

FIGURE 6.4 *Pattern of recruitment, by week*
Source: Survey of Central American Women in Houston.

that have previously received little research attention, and where there is limited experience with survey research. It is therefore important to be prepared to make adjustments throughout the process of data collection. In order to make the right adjustments, parallel monitoring during data collection process is crucial.

Methods for parallel monitoring

Effective recruitment requires closely observing the process to detect any potential problems or recruitment anomalies that will explain why recruitment is not occurring (e.g., bottlenecks, missing sub-groups, selling and bartering of coupons, etc.). This observation can be done using several data sources; for instance, by continuing communication with key informants and the target population, and recording any observations made by survey staff. In addition, using a coupon log of distributed and redeemed survey coupons will map out recruitment across specific variables to detect recruitment patterns, coupon numbering errors and missed sub-populations. In Figure 6.5, a coupon log was used to examine cross-site recruitment between two sites. In this example, only one chain was successful in recruiting across the two sites. Based on this diagram,

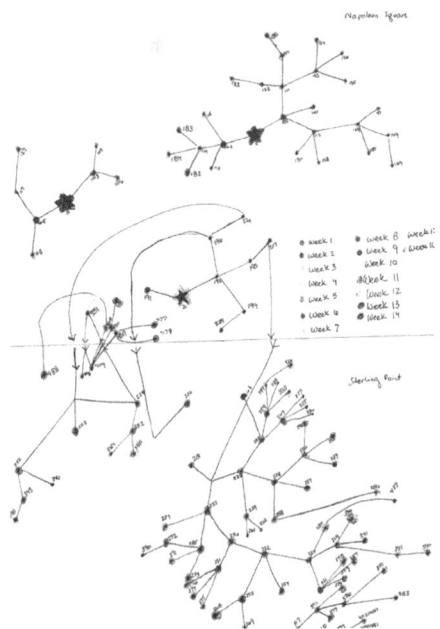

FIGURE 6.5 *Paper and pencil diagram to monitor cross-site recruitment*
Source: Survey of Central American Women in Houston.

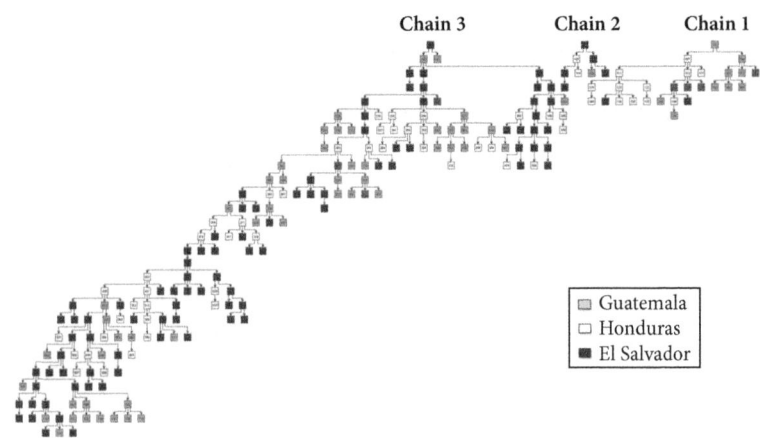

FIGURE 6.6 *Diagram of recruitment, by country of origin*
Note: Created in yED Graph Editor® software.
Source: Survey of Central American Women in Houston.

the researcher may be prompted to consider modifications to improve cross-site recruitment, including specifically asking respondents to recruit someone who would probably go to the other site. Note that these diagrams can be quite simple. A paper and pencil sketch can be updated with little effort and provides the research team with a real-time view of survey progress.

The same type of diagram can be used to monitor recruitment along key characteristics of respondents, especially if the investigator suspects that there are barriers to social mixing across groups. For example, in the *Central American Women in Houston* survey, researchers were interested in mapping recruitment by nationality because they suspected a high degree of homophily (the principle that contact between similar people occurs at a higher rate than between dissimilar people) among women from the same country of origin. Researchers periodically plotted recruitment between Guatemalans, Hondurans, and Salvadorans to help uncover potential bottlenecks across country of origin (Figure 6.6). These plots indicated no major bottlenecks among the sub-populations.

Many RDS surveys use computerized records to store recruitment information and monitor recruitment chains, but a paper and pencil backup is advisable to allow cross-checking in case of errors. In the *THEMIS-Norway* survey, the coupon logs were made accessible to team members using a web-based document. This was essential given that data collection and the disbursement of secondary incentives occurred at different sites. After conducting an interview, interviewers immediately updated the virtual document. A designated staff member was then able to administer secondary incentives.

Addressing slow recruitment

Slow recruitment may be a result of the selection and training of seeds. If the initial seeds are not motivated and well connected, they are unlikely to recruit quickly and effectively (see Chapter 4). If this is the case, finding other seeds may be sufficient to increase recruitment. Slow recruitment may also be related to inappropriate or too low incentives (see Chapter 5). An obvious solution may be to increase incentives, but it is useful to investigate other possible reasons for slow recruitment and evaluate the costs and benefits resulting from this change.

Slow recruitment could also be an indication that respondents do not trust the staff and the motives behind the survey. Trust is important in

RDS because respondents elect to enroll in the survey by presenting themselves at a fixed-site or calling for an appointment.

Many RDS surveys with migrant populations have had to address the issue of trust. In the *Polonia in Dublin* survey, some respondents expressed concerns with the possible links to government institutions and had to be reassured that their information would be kept confidential. In the *Nigerians in NYC* survey, the researcher worried that respondents would be suspicious about giving up economic information to a stranger. This was remedied by gaining the endorsement of religious leaders within the community who, during religious services, encouraged the community to participate in the survey.

Slow recruitment can be related to the choice of the survey site. As discussed in the case studies in the beginning of this chapter, the research team in the *THEMIS-Norway* surveys started with a fixed-site approach but moved to a mobile-site to be more flexible towards potential respondents' schedules and preferences. Allowing for both appointments and drop-in hours convenient to the respondents, as occurred in the *Central American Women in Houston* survey, may also improve slow recruitment. At the same time, slow recruitment may be due to very practical barriers, for example, individuals not being able to find the survey site, the presence of a security guard, or because of the location of the survey site (e.g., close to a police station or in a dangerous area).

Addressing rapid recruitment

Some surveys experience recruitment that progresses too rapidly. Uneven recruitment patterns are common and not problematic as long as survey staff can manage it. However, unmanageable recruitment flow requires a careful response. Figure 6.7 shows recruitment in the *Polonia in Oslo 2006* survey. After some weeks of slow recruitment, there was a rapid increase in recruitment to 50 respondents per week in week 6 and then to over 100 respondents per week in weeks 8 through 10. This rapid increase in recruitment was not problematic as this survey had sufficient staff and space, but without proper planning such an increase in recruitment could become unmanageable.

If survey staff are unable to interview all individuals as they arrive at the survey site, various strategies can be implemented to slow down recruitment to a more controllable pace. One strategy is to move from a drop-in to an appointment-based system if this is suitable for the target

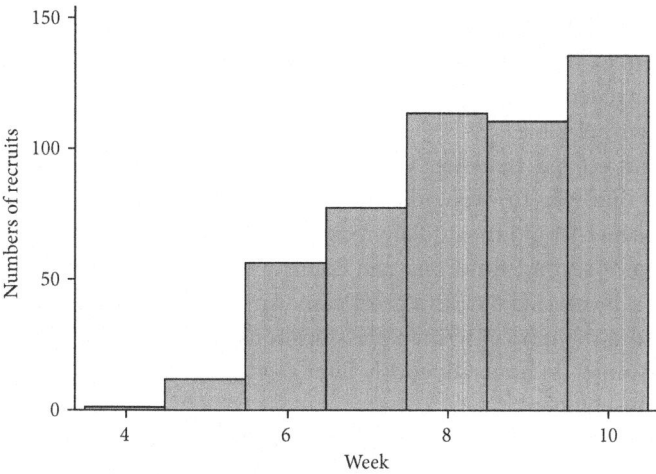

FIGURE 6.7 *Pattern of rapid recruitment, by week*
Source: Survey of Polonia in Oslo, 2006.

population. Another strategy is to use activation dates, whereby coupons are not immediately valid (Johnston, 2013b). Finally, researchers can reduce the number of coupons provided to respondents from three to two, and then from two to one. This strategy not only slows recruitments, but also allows for the generation of longer recruitment chains (Johnston et al., 2007; Johnston, 2013b).

Masquerading and repeat respondents

Two challenges to conducting RDS surveys are masquerading, whereby someone who is not part of the target population tries to participate, and repeat participation. Both masquerading and repeat participation could indicate that incentives are too high (see Chapter 5). In the *Polonia in Reykjavik* survey, individuals of other nationalities tried to participate in the survey, but were easily identified as non-Polish by the native speakers on the team. Using either staff from the same ethnic group, or others who can easily identify members of the target population, can reduce masquerading. For a survey that focuses on one specific linguistic group, masquerading may be relatively easy to identify by having interviewers or coordinators who speak the same language. However, it can be more complex to identify if the population is defined by length of stay in the country or other variables that have no objective criteria from which members of the population can be distinguished from non-members. Care is needed to identify masquerades, while not excluding those who are actually eligible.

Having a fixed screener, a person to "meet and greet" respondents, may be one way to overcome masquerading and repeat participation. The screener would most likely recognize a person who presents themselves for a second time, and can confirm eligibility. Screening should, however, be carried out with great care. Any eligibility screening should be based on empirical written classifications, to ensure that the screening personnel do not erroneously filter away parts of the population that should be eligible.

In appointment-based approaches, consider tracking phone numbers to avoid double entries. It should be pointed out that if logs of phone numbers are kept, full anonymity can no longer be promised, even if the information obtained is treated confidentially. Furthermore, screeners should be open to legitimate respondents sharing a phone if they do not have their own.

Ending RDS

Ending an RDS survey requires careful planning. Ideally, there should be no valid coupons in the community at the time the survey ends. Strategies include establishing a date to cease coupon distribution, reducing coupon expiration time, moving to an earlier expiration date, and reducing the number of coupons given to each respondent.

As the survey approaches the calculated sample size, it is necessary to consider when to cease coupon distribution. Use parallel monitoring to calculate the average number of respondents enrolling daily, in order to establish how many additional days are needed to attain the calculated sample size. Also, consider the number of valid coupons in the community and the percentage of coupons being redeemed to anticipate the expected number of remaining respondents.

In the *Polonia in Oslo* survey, researchers additionally made the decision to provide "closing incentives", whereby respondents received a slightly larger primary incentive to compensate them for not being asked to recruit.

In the *Central American Women in Houston* survey, the initial coupon expiration time of one month was consecutively reduced to two weeks, one week, three days, and finally one day. Additionally, the researchers stayed at the survey site for another week to honor respondents with valid coupons disbursed before the end date was decided. The *sub-Saharan Africans in Morocco* survey reduced coupons by reducing the number of coupons from three to two or two to one to gradually end the survey (Johnston, 2007; 2013b). Some studies have, on the other hand, abruptly stopped recruitment, even when individuals still had valid coupons, resulting in resentment in the community and damaging future research opportunities with the population.

Ethical considerations

In peer-to-peer recruitment, trust in the research and a positive interview experience are essential to the success of the survey. Thus, it is particularly important to pay attention to issues of anonymity and confidentiality, as well as to ensure that respondents are aware of these efforts. Particularly when using staff from the same community as the interviewees, who may know each other directly or indirectly, it is essential that all parties are informed about the procedures of ensuring confidentiality and anonymity. In the *THEMIS-Norway* surveys, for example, all staff members were asked to sign a "confidentiality agreement" with the institution responsible for the research, stating that the information they obtained through the project would not be discussed outside of the project team. Additionally, an explanation of the confidentiality procedures was read to each respondent at the beginning of the interview, and a copy signed by the interviewer was given to him/her to keep.

RDS studies often, though not always, target vulnerable or hidden populations, which brings additional challenges. Many of these challenges are present in other research methods for vulnerable and migrant populations (Van Liempt & Bilger, 2012) but it is worth reiterating that the protection of respondents needs to be ensured throughout the fieldwork. Researching vulnerable and/or migrant populations may mean that the research team members are the respondents' first contact with an institution in the host country. Team members should therefore be prepared for situations where respondents disclose problematic information or are seeking help. In most cases, this will mean providing information or referral to appropriate services. Indeed, one could make such information available not only on request, but also provide it at the survey site for all respondents.

A further concern is the safety of not only respondents, but also of staff, especially when using mobile sites as in the *Nigerians in NYC* and *THEMIS*-Norway surveys. Again, this is not unique to RDS surveys, as survey interviewers often visit respondents' homes, which in itself carries certain risks. However, RDS staff members need to handle money as part of the survey design, and this is likely to become known to respondents and others. Careful consideration in the protocol on how interviewers are assigned and how money is dispensed are therefore recommended.

Finally, every researcher will need to check the ethical review procedures of their own institution. As RDS studies may represent a relatively new and novel method in a given setting and are often used for vulnerable populations, additional considerations in this process are likely, and sufficient time should be allocated for this. Throughout the chapter, we have advocated that sometimes it is necessary to make changes to the survey design after fieldwork has started. Naturally, such changes need to be within the parameters of an ethical review, or may otherwise require an additional review. It is good practice to anticipate some of the potential changes needed and get approval for those in the initial review. An example of this foresight could be the setting of incentives within a certain acceptable range rather than a fixed amount, and allowing some flexibility in terms of fixed versus mobile sites.

Conclusion

To our knowledge, many RDS surveys of migrant populations have had to modify some design features during data collection. Even with detailed preparation, it is likely that future studies will encounter issues that cannot be anticipated. As long as the rigorous methodological principles of RDS are met, many modifications are possible during this stage. However, certain aspects of RDS data collection should not be modified once the survey has begun. This includes the eligibility criteria upon which the probability of selection through the personal network sizes is determined (see Chapter 3). Survey logistics, on the other hand, can and often need to be adapted to ensure the success of RDS recruitment.

It can be particularly difficult to plan an RDS survey for a population in which it has previously never been employed, as populations respond differently to RDS survey designs. Formative assessment and ongoing parallel monitoring are particularly important for identifying challenges and adjusting to the population as quickly and in the best manner possible.

7
Analyzing Data in RDS

Lisa G. Johnston and Renee Luthra

Abstract: *This chapter reviews how to analyze data gathered using RDS. We begin by reviewing why RDS data cannot be analyzed with general statistical software, and suggest several alternative software options designed for RDS data analysis. We briefly review the different estimators currently in use, and the estimation of variance in RDS analysis. We discuss potential sources of bias in RDS data, including seed dependence, homophily, differential recruitment, and bottlenecks, describing how to diagnose these problems during analysis. In addition, we describe the common practice of exporting weights for multivariate analysis using RDS data. Finally, we discuss the responsible reporting of results from RDS data analysis and provide examples of the use of RDS data to impact policy.*

Tyldum, Guri and Lisa G. Johnston, eds. *Applying Respondent Driven Sampling to Migrant Populations: Lessons from the Field.* Basingstoke: Palgrave Macmillan, 2014. DOI: 10.1057/9781137363619.0014.

Introduction

Respondent Driven Sampling (RDS) is a methodology for both collecting and analyzing data. If data collected through RDS methods are not analyzed to correct for specific biases, the survey should not be referred to as "RDS" (unless it can be shown that the data is completely self-weighting, that is, the adjusted estimates are the same as the unadjusted statistics). Many researchers who have tried RDS for the first time have been surprised to learn that once data is collected, it cannot simply be imported into a standard statistics package for analysis. To analyze RDS data, it is necessary to use one of the specialized software programs that generate estimators and confidence bounds specific to RDS assumptions (see Chapter 1 for a presentation of RDS assumptions).

The unit of analysis in RDS is a network structure rather than an individual, and the analysis generalizes to the networks of the sampled population. This method of analysis influences the kind of estimators we use, the way we understand the variance around the estimates, and how we interpret the findings. This chapter reviews some of the most important concepts and approaches used to analyze RDS data, responding particularly to the following questions: Why do I need to conduct special analysis with RDS data? What are the software options available for RDS analysis and where do I find them? How do the different estimators perform under different sampling conditions, and how do I decide which one to use? Can I calculate variance for the estimates? How do I diagnose and reduce bias in my sample? Is it possible to conduct multivariate analysis with RDS data? And, how should RDS findings be reported? Throughout, and at the end of the book, are references useful for finding more information about RDS data analysis and links to data analysis tools.

A need for special analysis of RDS data

Unadjusted statistics from a sample can be used to describe the sample, but unless the sample is drawn in a way to produce representative data, such statistics can only describe the sample. For instance, a hypothetical convenience survey using snowball sampling of Haitian migrants in the Dominican Republic would result in estimates that represent only those migrants who ended up in the sample, rather than the population of Haitian migrants. In snowball sampling the probability of inclusion is generally unknown and samples tend to over-represent individuals with more peers (potential recruits or referrals) and under-represent individuals with few peers. If individuals with many peers are more likely to be, for instance, unemployed, male or uneducated, then the prevalence of these characteristics will be overestimated. These findings cannot be extrapolated to the wider population of Haitian migrants as they only represent the *sample* of Haitian migrants.

RDS is similar to a snowball sampling method but incorporates numerous methodological and statistical elements to mitigate the biases in snowball sampling (Heckathorn, 2002; 2007; Salganik & Heckathorn, 2004). At the analysis stage, data must be weighted to account for the over-representation of people with many peers and underestimation of people with few peers. RDS population estimates will generalize to the *network of the population* from which the sample was drawn, which is essentially representative of the *population* from which it was drawn, if all RDS assumptions are met. The example in Table 7.1 is taken from the survey among Francophone *sub-Saharan Africans in Morocco* (see Appendix I, for a thorough presentation of all surveys referred to in this volume). The sample statistics show that there are 67.3% males and 32.7% females in the sample. The population estimates, however, adjusted to account for biases of under and over-representation show that there are 63.9% males and 36.1% females. If we had not adjusted our data, we would have overestimated males and underestimated females by 3.4% (see Table 7.1).

Which software to use when analyzing RDS data

Fortunately, there are free software packages available for making adjustments to data collected with RDS. Currently there are two widely used

TABLE 7.1 *Sample and population estimates of gender distribution*

	Sample estimate (%)	Population estimate (%)	Confidence intervals	Standard error	Sample sizes
Male	67.3	63.9	57.7, 70.1	0.032	276
Female	32.7	36.1	29.9, 42.3	0.023	134

Source: Survey of francophone sub-Saharan migrants in Morocco.

software programs: the RDS Analysis Tool (RDSAT) and RDS-Analyst. There is also an add-on package to STATA which allows RDS data to be analyzed with the RDS-I estimator (Schonlau & Liebau 2010). The current version of RDSAT at the time of publication is 7.1 and is available for free from www.respondentdrivensampling.org. RDSAT is widely used and now has features for analyzing several databases at once and for saving workspaces. However, the estimators currently available in RDSAT are limited to RDS I, an early RDS estimator highly dependent on the RDS statistical assumptions (see section below for more information on the different estimators).

RDS-Analyst was developed by a group of statisticians and researchers as a more user-friendly alternative to RDSAT and can be downloaded for free at (www.HPMRG.org). It is based in R Project for Statistical Computing (a free software programming language); has graphical user interface with drop down boxes, which makes analysis easier; includes all the current estimators (estimators available up to 2013); and allows direct downloading of all file types (SPSS, STATA, SAS, R, Excel, txt, etc.). In addition, it allows you to build graphics and plots to use in diagnosing bottlenecks, convergence (or equilibrium) and other biases in the data, as well as displaying results. In addition, a command line in the package that allows the user to reproduce analysis.

Deciding which estimator to use

Current estimators for RDS analysis are primarily developed to describe proportions in a network and make inference about an entire network based on information about the known part of the network (the part in your sample). The evolution of RDS estimators since 1997 is based on improvements in our understanding of RDS assumptions and on how to match those assumptions with statistical techniques. The original paper on RDS methodology (Heckathorn, 1997) demonstrates that sample proportions are representative of the population proportions if the sampling procedures meet all the assumptions of the *random walk, first order Markov process model*.[1] Heckathorn (2002) later used differences in network sizes and homophily (the principle that contact between similar people occurs at a higher rate than between dissimilar people) across groups to adjust estimates and *incorporated data smoothing*[2] to control for the differential recruitment that frequently occurs in RDS samples. The RDS-I estimator, developed by Salganik and Heckathorn (2004) proposed adjustments based on transition probabilities of a recruitment matrix of who recruited whom, to account for differential recruitment (i.e., the probability that a person with a particular characteristic will recruit a person with the same characteristic) and self-reported network sizes. Because differential recruitment is measured for each individual variable, the weights applied to any given variable are not the same as those applied to other variables. RDS-I works well when homophily and differential recruitment exist (Gile et al., 2014). The RDS-II estimator (Volz & Heckathorn, 2008) makes use of network sizes to adjust estimates, and allows for weights to be applied to the entire sample, rather than to each variable separately. All the above-mentioned estimators rely heavily on the assumptions of the *random walk, first order Markov process model* (Goel & Salganik, 2009; Gile & Handcock, 2010).

The more recent Successive Sampling (SS) estimator (Gile, 2011) does not rely on meeting all of the random walk, first order Markov process model assumptions; specifically the assumption of *with-replacement sampling*. Instead, the SS estimator requires some knowledge about the size of the target population (population count). There will often be parameters on which to estimate the size of the target population, even if an accurate count is unavailable. In estimating the population count for the SS estimator, it is better to err on the high side.

In most situations (short recruitment chains [seed dependence], high homophily and *high referral bias* [one group referred more often than another group]) RDS-II outperforms RDS-I, and in most of these similar situations the SS estimator outperforms RDS-II (Gile & Handcock, 2010). If the sampling fraction is 30% or more, the SS estimator is recommended (Gile & Handcock, 2010; Gile, 2011). Appendix II provides a list of all estimators, the data required, limitations, type of variance estimation, analysis feature, and states in which software it is available.

Table 7.2 shows an example of output from the RDS-Analyst, displaying point estimates, confidence intervals, design effects, standard errors, and sample size from an analysis of males taken from the survey of *sub-Saharan Africans in Morocco*. First, look at the RDS-I and RDS-II estimates. They are different by about six percentage points, and have higher design effects than the SS estimators (indicating that the sample size was insufficient, since the sample size calculation for this survey used a design effect of 2).

In this example, the sample size is 276 and the true population count is about 4,000, giving a small sample fraction of about 7%. For this reason, the RDS-II and the SS estimator (based on the true population count) perform similarly, estimating that 63% of the population is male. When the estimated population count is less than the true population count, the SS estimator may perform poorly (closer to the RDS-I estimator, see underestimation of population count: 500 in Table 7.2) and have a lower standard error, whereas when the population count is overestimated (see overestimation of population count: 10,000 in Table 7.2), the estimator is similar to the RDS-II estimator and SS estimator using the true population size.

Note that the extent to which the various estimators are similar will not necessarily be the same for all data, as this will depend on the population structure (e.g., the existence of homophily, differential recruitment activity, bottlenecks, etc.), sampling biases (e.g., reciprocal recruitment relationships, accurate measurement of each respondent's network size, etc.), the quality of data collection (e.g., repetitive enrollment, enrollment of those masquerading as being eligible, bartering and selling of coupons, etc.), and the properties of the sample (e.g., attainment of long recruitment chains made up of numerous waves, diverse seeds, etc.) (Gile et al., 2014).

TABLE 7.2 *Example of output from RDS-Analyst*

	Point estimate	95% Lower bound	95% Upper bound	Estimated Design effect	Standard error	Sample size
RDS-I	0.6920	0.6166	0.7674	3.58	0.0385	276
RDS-II	0.6342	0.5559	0.7125	3.55	0.0400	276
SS estimator (true population 4000)	0.6389	0.5498	0.7280	2.62	0.0455	276
SS estimator (underestimation of population 500)	0.6623	0.6329	0.6916	2.28	0.0150	276
SS estimator (overestimation of population 10,000)	0.6352	0.5656	0.7047	2.32	0.0355	276

Source: Survey of francophone sub-Saharan Africans in Morocco.

Variance in RDS analysis

In RDS, as in other probability-based survey methodologies, we use the variance in our dataset to estimate the variance in the population, and based on this, we can evaluate the level of an estimator's precision. The larger the variance, the larger sample size is needed for a precise estimate. Most of the confidence bounds in RDS analysis (with the exception of RDS-II) are calculated using some variation of a bootstrap method (Efron & Tibshirani, 1986). This method is implemented by constructing numerous resamples (of equal size) of the observed dataset, providing us with the confidence bounds around that estimator. In bootstrap methods the final estimates and bounds may change slightly (usually no more than .05%) for the same variable with each analysis.

Assessing bias in RDS analysis

In RDS analysis there are several ways to assess the level of bias. Bias in RDS data is specific to the variables analyzed; there may be a greater bias in one variable compared to another variable in the sample. For example, while the bias in the variables of gender or age might be large, the bias in the variable of education might be small. Below we discuss four potential sources of bias in RDS: seed dependence, the degree of homophily, differential recruitment activity and bottlenecks.

Seed dependence

Sampling in RDS starts with purposively selected seeds, which may or may not have characteristics that represent the underlying network structure of the population. Individuals tend to be similar to the others in their social networks in a number of characteristics, such as educational level, place of residence and political preferences. Therefore, if the sample does not reach all the sub-populations in a network, which is likely if recruitment chains are short, then the sample may represent the characteristics of the seeds rather than that of the population. One way to determine whether the final sample is dependent on the seeds is to measure whether equilibrium or convergence has been achieved in the sample. Equilibrium is assessed by determining the sample proportion (i.e., by dividing the number of persons with or without a characteristic over the total number of people in the sample) at each successive wave. At some point (at a particular wave), the sample proportion will no longer change from one wave to another. This point of equilibrium indicates that the sample has started to represent a random mix of characteristics upon which the population is structured. The attainment of equilibrium is NOT the point at which to stop sampling. Rather, it is necessary to recruit many waves beyond the point of equilibrium to ensure that the equilibrium of proportions remains stable over numerous successive waves, and to attain an adequate sample size not marked by seed dependence. The point at which equilibrium is attained can be different for each variable in the same sample. In the first graph in Figure 7.1 (self-reported

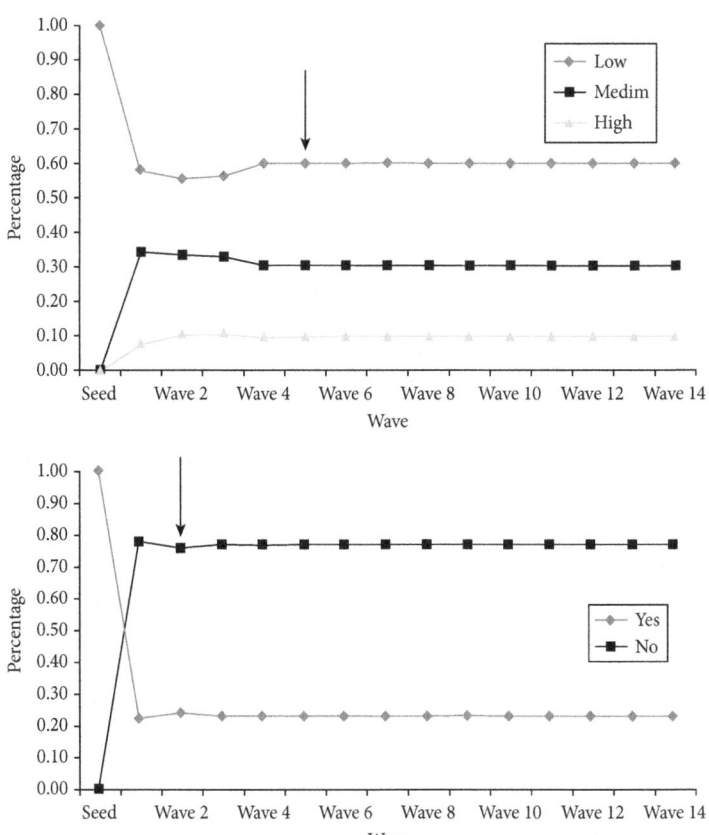

FIGURE 7.1 *Equilibrium points for two variables*
Source: Survey of Francophone sub-Saharan Africans in Morocco.

health status), equilibrium appears to be attained at wave 5 (see arrow), whereas in the second graph (service utilization), equilibrium appears to be attained at wave 2.

When equilibrium is not attained, it may be an indication that sampling needs to continue in order to add more waves to the sample. Identifying bottlenecks in the sample population (see Chapter 2), strategically selecting seeds to increase the opportunity to attain a diverse mix of respondents (see Chapter 4), and encouraging respondents to recruit members randomly from their network (see Chapters 4 and 6), will help in reaching equilibrium early on in recruitment (Johnston, 2013b). It is important to note that attaining equilibrium is not an indication that the sample is completely free from bias, as other factors are also involved in making up a good sample, such as achieving the calculated sample size. It is possible to attain equilibrium for each variable but still have bias from other sources.

Homophily

Two types of homophily can bias an RDS sample: population homophily and recruitment homophily. Population homophily refers to the social ties in the population sampled, whereas recruitment homophily refers to the ties in the recruitment chain. By social ties, we mean two people who are tied in a social network in the population, also known as a "couple". Population homophily is calculated as the ratio of the expected number of couples who share the characteristic of interest (e.g., both are female) if distribution was random (relative to the actual number of couples who share the same characteristic). Hence, values of population homophily greater than one indicate that more than the expected number of couples in the population are affiliated, based on the characteristics of interest, while a value of one means that ties are random. A value less than one indicates heterophily, meaning that there are more discordant ties than expected due to chance. For instance, if population homophily for sex is 0.75, there are 25% more sex-discordant couples than expected due to chance; if population homophily on sex is 1.1 (as displayed in Figure 7.2, below), there are 10% more same-sex couples than expected if recruitment was random.

Recruitment homophily is calculated as the homophily in regard to a given characteristic in the recruitment chains. It is calculated as the ratio of the number of recruits who share the same characteristic as their recruiter; relative to the number we would expect if distribution was random. Figure 7.2 illustrates the recruitment and population homophily

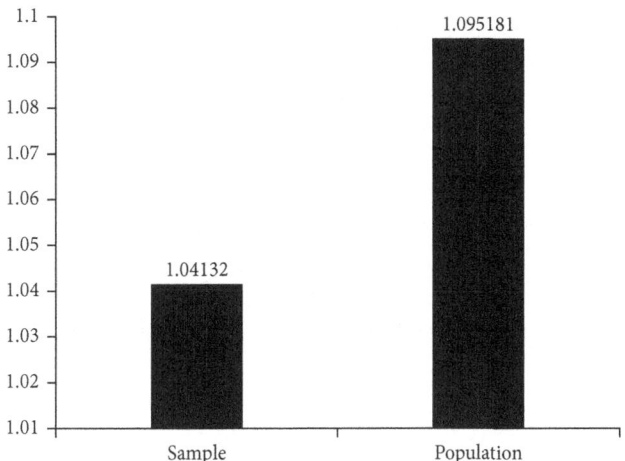

FIGURE 7.2 *Homophily for females in sample and population*
Source: Survey of francophone sub-Saharan Africans in Morocco.

levels for gender in the survey among francophone sub-Saharan Africans in Morocco. In this example, both recruitment and population homophily are low.

All RDS estimators adjust for some level of recruitment homophily; however, high recruitment homophily can be an indication of seed bias (i.e., that equilibrium has not been attained) and/or that recruitment is stuck in one sub-group (e.g., only males are represented in a sample of males and females). High recruitment or population homophily will result in unstable estimates and larger variance. High homophily has been found in several studies of migrants including the *Central American Women in Houston* survey (Montealegre et al., 2011), where substantial homophily was found by country of origin (e.g., Guatemalans vs. El Salvadorans). In the *Polonia in Oslo* survey, homophily was found between workers who were working for Polish firms versus those employed in Norwegian companies, and between those working for private households versus those employed by a firm. During the planning stages of the *sub-Saharan Africans in Morocco* survey, it was predicted that there would be high homophily between francophone and anglophone migrants so the researchers decided to have two separate samples based on language rather than just one sample regardless of language (Johnston, 2013a).

Differential recruitment activity

Differential recruitment activity is an indication of the relative connectedness of one group to another. It is measured as the ratio of the mean personal social network sizes of one group (e.g., females) relative to the mean personal social network sizes of another group (e.g., males). Differential recruitment results in under- or over-representation of some groups in an RDS sample. For example, if we are interested in knowing the proportion of females in the population, and females have a higher average number of connections in the social network than males, then females may be under-represented in the estimate. Under most circumstances, the RDS estimators are designed to adjust for differential recruitment activity (Gile & Handcock, 2010).

Analyzing bottlenecks

Bottlenecks can be identified using a graphic network package such as UCINET NetDraw (http://www.analytictech.com/downloadnd.htm), Gelphi (https://gephi.org/), Pajek (http://vlado.fmf.uni-lj.si/pub/networks/pajek/) or the diagnostics program in RDS-Analyst (www.hpmrg.org) (see more on bottlenecks in Chapter 2. Figure 7.3 displays a recruitment graphic made with UCINET NetDraw of the distribution of two neighborhoods in a mock sample. The grey nodes represent Neighborhood 1 and the black nodes, Neighborhood 2. Ideally, in a complete network component, the black and gray nodes should be interspersed to reflect recruitment across neighborhoods. However, in this case, all of the respondents are clustered in either one of the two neighborhoods, forming multiple isolated components. This situation indicates a bottleneck or a structural or social barrier that impedes affiliation and/or recruitment. A similar bottleneck occurred in a study of Somalis in Oslo (Gele et al., 2012) whereby four seeds resulted in four separate samples, as young and old did not recruit across their sub-groups and men and women did not recruit across their sub-groups.

Ideally, bottlenecks should be identified before data collection starts, as they are easier to reduce at this stage (through seed selection [see Chapter 4], stratified secondary incentives [see Chapter 5], and by motivating

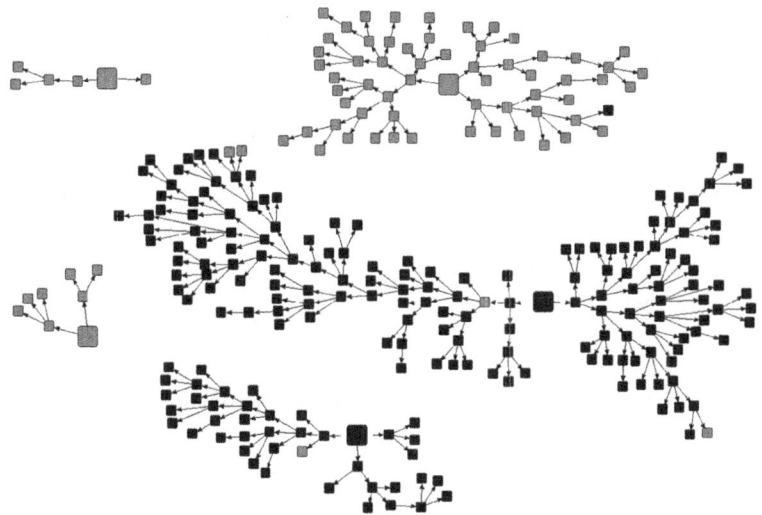

FIGURE 7.3   *Bottlenecks within neighborhoods*
Legend: Gray = neighborhood 1, Black = neighborhood 2.
Source: Authors' simulation.

respondents to recruit across [see Chapters 4 and 6]). However, if bottlenecks are found at the analysis stage, important decisions about how to analyze the data may be required (Gile et al., 2014). In the worst case, such as with the recruitment structure seen in Figure 7.3, survey data may have to be analyzed as two distinct samples. Luckily, in this case, the two neighborhoods were bottlenecked almost completely (i.e., few to no grey nodes in the black node chains and vice versa), so that the sample could be split across neighborhoods, and analyzed separately. If there were more gray nodes in the black clusters and vice versa, but not in sufficient amounts to be considered one complete network, it would be difficult to split the sample into two. Keep in mind that splitting the sample results into two samples of smaller size will likely result in a lower power and confidence than that calculated in the initial sample size. An alternative is to analyze several important variables for each of the bottlenecked sub-groups (for each of the two neighborhoods in the above example) to see if there are differences. If similar estimates come from the two bottlenecked sub-groups, then data can possibly be analyzed together.

While social and structural barriers that impede affiliation and/or recruitment are present in all populations, migrants may be more susceptible to specific types of bottlenecks. For instance, common characteristics that define connections particular to migration research include the workplace and neighborhood of residence (McPherson et al., 2001), employment and access to civil life, country of origin, and language. Additionally, information sharing about resources and local organizations commonly occurs through word of mouth in many migrant communities (Kasinitz, 2008; Portes & Rumbaut, 2001; Drever & Hoffmeister, 2008) leaving migrant populations susceptible to physical barriers, such as distance and availability of transport.

Exporting weights for multivariate analysis

The central research questions of many migration scholars using RDS to sample migrant populations, including sociologists, anthropologists, and demographers, are multivariate in nature. Migration scholars want to understand what impacts decisions to migrate, attitudes to return, and successful integration. While bivariate analysis can go a long way, complex relationships and interdependencies often demand multivariate models.

Unfortunately, the techniques available to perform regression or other more advanced analysis on a network are complicated and not yet sufficiently developed. Currently, it is proposed that researchers export weights from existing RDS analysis programs (both RDSAT and RDS-Analyst allow you to export weights) to adjust for some biases. However, this level of analysis should be interpreted with caution, as the weights do not account for several aspects of network sampling and bootstrap variance. Additionally, the types of weights that are exported vary, depending on the estimator used. The RDS-I estimation techniques use weights that are based on the average network sizes in the categories of the variable analyzed and are therefore variable dependent. In contrast, the RDS-II and SS estimators use weights based on the network characteristics of the individuals, and are not variable dependent.

One suggestion when conducting multivariate analysis is to conduct the analysis with weights and then repeat it without the weights. If the results of the two analyses are the same, then report the unweighted result (Heckathorn, 2007 citing Winship & Radbill, 1994), as weights introduce additional variance that impact confidence intervals.

Reporting RDS findings

It is often not until the data cleaning and analysis stage that biases in the sampling method are found. It is also common that mistakes or unforeseen events occur during the data collection and analysis phases. Thus, it is essential to report anything that can potentially impact the interpretation of findings (Malekinejad et al., 2008; White et al., 2012; Gile et al., 2014). To assess the quality of the data collection, a reader would need to know the number of seeds used in recruitment, the number of recruitment coupons per recruit, and the number, type, and location of the recruitment venue(s). Also relevant is a discussion of how the survey organization may have created bias in the data. For example, in a survey of migrants who are not legally residing in a country, conducting a survey in a government institution may dissuade some migrants from participating. Other sources of biases to report are any identified selling, bartering or other misuse of coupons, and the distribution of coupons to strangers and people waiting in front of the interview site. Given that the social network size is essential to the analysis of RDS data, the social network size question(s) used to measure these networks should also be provided in any reporting. The number of recruits and maximum number of waves produced by each seed can also be useful for understanding the sample (see Table 7.3).

In order for the reader to assess the quality of the data analysis, the presentation of survey results should describe all statistical methods, including the name of the estimator used to take account of the RDS strategy, as well as the software package and version number. Finally, when discussing the results, consider the limitations of the RDS sampling

TABLE 7.3 *Description of recruitment by seeds*

	Number of recruits*	Maximum number of waves*	% of sample
Seed 1	5	3	2.2
Seed 2	158	14	57.4
Seed 3	7	2	2.9
Seed 4	61	6	22.4
Seed 5	41	7	15.2

Note: *Excluding seeds. Authors' simulation.

method and, if used, the RDS method(s) of inference, and possibly comment on how representative the unadjusted sample is thought to be. In the end, report everything that would enable any reader to properly evaluate the RDS methodology used, and the level of representativeness of the analysis involved.

Using RDS findings to impact policy

RDS has been used worldwide in hundreds of surveys for policy and health-related research. As a sampling strategy that can yield more rigorous and representative disease prevalence estimates, RDS has been the preferred sampling method for numerous biological and behavioral surveillance surveys of non-migrant populations at high risk for HIV exposure (Malekinejad et al., 2008). Findings from these surveys provide much-needed data for allocating funding, evaluating program success, and for planning any necessary intervention and prevention programs. Not until recently has RDS been used among migrants to measure disease prevalence and health seeking behaviors. Specifically, RDS has been used to measure healthcare coverage and service access among migrant communities in the United States (Montealegre et al., 2012; Montealegre et al., 2011), malaria prevalence among cross-border working migrants in Thailand (Khamsiriwatchara et al., 2011),and the utilization of HIV counseling and testing in India (Solomon et al., 2013). Findings from the *sub-Saharan Africans in Morocco* survey, which collected data on HIV and syphilis prevalence, migration experience, health access, and stigma and discrimination, were disseminated during a two-day meeting to ministers and other government officials, to NGOs working with migrants, and to donor agencies and other stake holders. These findings will be used to inform Morocco's new asylum and migration policies (Johnston, 2013a).

Other surveys using RDS among migrants provide data that allows for a closer understanding of the nature of migrants residing in foreign countries, their living and working standards, quality of life, and access to services. For instance, the *Polonia in Oslo* surveys were able to demonstrate that a substantial group of Polish migrants in Oslo worked in the shadow economy – some of them in exploitative and illegal working conditions – and the data was important in formulating subsequent government policies. Another example is an RDS survey of low-wage workers, many of whom were undocumented migrants in New York, Los Angeles, and Chicago, which found high rates of employment and labor law violations (Bernhardt et al., 2009). Findings from this survey were featured in several national media sources, such as the New York Times, and drew comment from national labor organizations, as well as from the United States labor secretary.

Conclusion

The goal of this chapter was to outline some of the most common methods used for analyzing and assessing bias in RDS data. We believe that the benefits of RDS for studying migrant populations outweigh the drawbacks, insofar as we understand the biases in our samples and are transparent about these biases when publishing data. In this chapter we have outlined methods to identify bias after fieldwork has ended, and have described the different estimators, as well as looking at how to correct for some sources of bias, such as homophily and differential recruitment. The software programs that utilize these different estimators are described; all of these can be downloaded for free, and additional information on their use is available online. Finally, this chapter has focused on the many uses of RDS data in analysis, and discussed some of ways RDS data of migrant populations have been analyzed. As the popularity of using RDS to sample migrant populations continues to grow, we expect that the type and use of these analyses, for both health monitoring and for more general integration purposes, will expand.

Notes

1 The random walk, first order Markov process model assumes that the population is joined by an informal social network of relationships. The model beings with an individual who is linked to one person; this process is extended as more individuals are linked. However, in a true random walk, links can go forwards and backwards (back to an already selected link), which is not realistic in RDS sampling. To overcome potential bias of links going forwards and backwards (also known as "with-replacement sampling") the population being sampled must have a small sampling fraction, which is sometimes

unrealistic. In addition, the random walk is a single non-branching linking process, which RDS is not. Finally, the random walk assumes that the sample will reach equilibrium (whereby the sample is independent with respect to characteristics to the non-randomly selected seeds), which may be difficult to attain in some populations.

2 Incorporated data smoothing is a process whereby the number of cross-group recruitments is averaged so that the recruitment matrix of who recruited whom is symmetric (see Heckathorn, 2002).

Appendix I: Summary of RDS Surveys Referenced

Central American Women in Houston

Proyecto Enlaces was an epidemiological study that aimed to explore the effectiveness of RDS in recruiting members of the undocumented Central American immigrant population, and to estimate the prevalence of HIV risk and testing behaviors. In 2010, RDS was used over the course of 16 weeks to recruit a sample of 226 Guatemalan, Honduran, and Salvadoran women living in Houston, Texas, USA, without a valid visa or valid residency papers. The researchers found that social networks were dense, that respondents adopted the recruitment system with reasonable ease, and that cross-group recruitment across sub-groups was sufficient to achieve a diverse sample that attained equilibrium for all demographic and sexual behavior characteristics (Montealegre et al., 2013). In regard to HIV risk and testing behaviors, the researchers found that recent immigrants have less stable sexual partnerships than established immigrants (Montealegre et al., 2012a), and that the relatively high prevalence of lifetime HIV testing (67%) was primarily due to routine testing as part of prenatal care (Montealegre et al., 2012b). The survey also found low levels of healthcare coverage (35%) and of access to a usual source of care (43%). Healthcare coverage among undocumented Central American immigrant women was primarily through the local indigent healthcare program and most of those with a usual formal source of care received care at a public healthcare clinic (Montealegre & Selwyn, 2014).

English language publications

Montealegre, J. R., J. M. Risser, K. Sabin, B. J. Selwyn, & S. A. McCurdy. 2013. "Effectiveness of Respondent Driven Sampling among Undocumented Central American Immigrant Women in Houston, Texas", *AIDS and Behavior*, (17): 719–727.

Montealegre, J. R., J. M. Risser, K. Sabin, B. J. Selwyn, & S. A. McCurdy. 2012. "Prevalence of HIV Risk Behaviors among Undocumented Central American Immigrant Women in Houston, Texas", *AIDS and Behavior*, 16 (6): 1641–1648.

Montealegre, J. R., J. M. Risser, B. J. Selwyn, K. Sabin, & S. A. McCurdy. 2012. "HIV Testing Behaviors among Undocumented Central American Immigrant Women in Houston, Texas", *Journal of Immigrant and Minority Health*, 14(1): 116–123.

Montealegre, J. R., & B. J. Selwyn. 2014. "Healthcare Coverage and Use Among Undocumented Central American Immigrant Women in Houston, Texas", *Journal of Immigrant and Minority Health*, 16(2): 204–210.

Foreign migrants in Ukraine

The RDS survey among Foreign Migrants in Ukraine was part of an HIV bio-behavioral surveillance survey aimed at estimating HIV prevalence and related risk behaviors among foreign migrants. The survey was conducted in 2013 in five cities of Ukraine (Kyiv, Donetsk, Odessa, Kharkiv, and Luhansk) by the International HIV/AIDS Alliance in the framework of the Global Fund to fight AIDS, Tuberculosis and Malaria project. Of the 1,100 foreign migrants sampled, 400 were labor migrants, 400 were students, and 300 were refugees or asylum seekers. Data collection took place over the course of between 8 and 12 weeks.

Migrants in Warsaw, 2010 and 2012

The 2012 survey of Ukrainians, Russians and Belarusians in Warsaw, Poland, was financed by the Polish National Bank (NBP) and conducted by experts from the Centre of Migration Research Foundation under the supervision of the NBP's Department of Statistics. The aim of the project was to test the methodological solutions, which in the future could be implemented in research on the situation of selected immigrant groups in Poland. Only immigrants who had been employed during the 12 months preceding the survey were eligible to participate, and a total of 548 migrants were interviewed. Particular emphasis was placed on the economic dimensions of the immigration, especially economic integration and the patterns of remittance behavior. Data was collected with the use of two methods – quota sampling and RDS, in order to allow researchers to assess their effectiveness. The results of this comparison showed that RDS was more effective and reliable than quota sampling, especially while studying sensitive topics.

The 2010 RDS study in Warsaw was conducted by the Centre of Migration Research, University of Warsaw, and was part of a larger project titled "Mobility and Migrations at the Time of Transformation – Methodological Challenges", which was also coordinated by this institution. This study was the first attempt to research immigrant groups in Poland with RDS. Eligible participants were those who currently lived in Warsaw or in the greater area of Warsaw, who came to Poland after 1989, and before leaving the country of origin held Ukrainian, Belarusian or Russian citizenship. Some 511 migrants were interviewed. The main purpose of this study was to check the functionality of RDS in researching this population, and possibly to research unregistered immigrants at a later date. The results of the study showed a lack of connections between two groups of immigrants – students and workers (a two-component network). The application of RDS enabled interviews with participants that worked in the shadow economy.

English language publications

Napierala, J., & A. Gorny. 2013. "Assessment of Effectiveness of RDS Sampling Method in Migration Studies". Paper presented during THEMIS project conference, *Examining Migration Dynamics: Networks and Beyond*. University of Oxford, September 24–26, 2013.

Publication in Polish

Napierala, J., & A. Górny. 2011. "Badania migrantów jako przedstawicieli populacji 'ukrytych'-dobór próby sterowany przez respondentów (Respondent Driven Sampling)", in P. Kaczmarczyk (ed.),

Mobilność i migracje w dobie transformacji, wyzwania metodologiczne.
Wydawnictwo Naukowe Scholar. Warszawa. S. 155–194.

Nigerians in New York City

The year-long Nigerians in New York City study was conducted in 2007, and consisted of a mixed methods approach in order to understand the economic incorporation of this population. While the specific interest of the study centered on the self-employed, the survey itself was conducted with both self-employed and waged laborers. The survey area covered the five boroughs of the city (Manhattan, Brooklyn, Bronx, Queens and Staten Island) and sampled 81 respondents. Findings contradict the "disadvantage hypothesis", which suggests that immigrants decide to become self-employed when they encounter barriers (language, legal status, human capital and other) to waged labor. For Nigerians, self-employment is a traditional and familiar occupational niche, and something that even those successful in the waged labor market often strive for. The project was financed by the United States National Science Foundation (NSF), as well as the Africana Research Center, the Research and Graduate Studies Office, the Population Research Institute, and the Anthropology Department at the Pennsylvania State University.

Polonia in Oslo, 2006 and 2010

The first Polonia survey was conducted as a pilot study in Oslo, Norway, in 2006 by the Fafo Institute for Labour and Social Research with assistance from the Centre of Migration Research in Warsaw and funding from the Norwegian Research Council and the Norwegian Ministry of Labour. The survey gathered data on the socio-demographic characteristics and migration histories, as well as the working and living conditions of Polish migrants in Oslo, and the survey results were published in a report the following year (Friberg & Tyldum, 2007). In 2010 the study was replicated with funding from EEA Grants – as part of the project "Mobility and Migrations at the Time of Transformation – Methodological Challenges", coordinated by the Centre of Migration Research at the University of Warsaw – with additional funding from the Norwegian Ministry of Labour. Again, the main results were published in a report the following year (Friberg & Eldring, 2011). Both times, more than 500 Polish migrants were interviewed face-to-face by a team of Polish-speaking interviewers. On average, each interview took about 45 minutes to complete.

English language publications

Friberg, J., J. Arnholtz, L. Eldring, N. Hansen, & F. Thorarins. /2014. "Nordic Labour Market Institutions and New Migrant Workers: Polish Migrants in Oslo, Copenhagen and Reykjavik", *European Journal of Industrial Relations*, 20(1): 37–53.

Friberg, J. H., & L. Eldring, (eds). 2013. "Labour Migrants from Central and Eastern Europe in the Nordic Countries – Patterns of Migration, Working Conditions and Recruitment Practices", *TemaNord*, 2013: 570.

Friberg, J. H. 2013. "The Polish Worker in Norway. Emerging Patterns of Migration, Employment and Incorporation after EU's Eastern Enlargement", PhD Dissertation. Fafo-report, 2013: 06.

Friberg, J. H. 2012. "The Stages of Migration. from Going Abroad to Settling Down: Post-Accession Polish Migrant Workers in Norway", *Journal of Ethnic and Migration Studies*, 38 (10): 1589–1605.

Friberg, J. H. 2012. "Culture at Work: Polish Migrants in the Ethnic Division of Labour on Norwegian Construction Sites", *Ethnic and Racial Studies*, 35 (11).

Friberg, J. H. 2012. "The Guest-Worker Syndrome Revisited? Migration and Employment Among Polish Workers in Norway's Capital", *Nordic Journal of Migration Research*, 2 (4).

Friberg, J. H., K. Tronstad, & J. E. Dølvik. 2012. "Central and Eastern European Labour Migration to Norway. Trends, Conditions and Challenges", in *Free Movement of Workers and Labour Market Adjustment, Recent experiences from OECD countries and the European Union.* OECD.

Friberg, J. H. 2010. "Working Conditions for Polish Construction Workers and Domestic Cleaners in Oslo: Segmentation, Inclusion and the Role of Policy", in Richard Black, Godfried Engbersen, Marek Okólski, & Cristina Pantîru (eds), *A Continent Moving West? EU Enlargement and Labor Migration from Central and Eastern Europe.* Amsterdam: Amsterdam University Press.

Polonia in Reykjavik

Polonia in Reykjavik, Iceland, was an RDS study conducted by the Center for Immigration Research at the Reykjavik Academy in 2010, targeting Polish migrants residing in the capital of Iceland. After two-and-a-half months a sample of 480 individuals was reached. The research was a part of the larger project "Mobility and Migrations at the Time of Transformation-Methodological Challenges", coordinated by Centre of Migration Research at the University of Warsaw. The main purpose of this project was to compare and evaluate different methodologies applied to survey migrant populations, including RDS. The questionnaire used covered a broad range of questions concerning issues such as the reason for migration, the pre-migration situation, the position of migrants in the labor market, and the family status of the migrants. The survey was the first extensive study of Polish migrants in Reykjavik, giving a good overview of the migrants' demographic characteristics, and the main patterns dominating this migration stream. The recruitment was relatively smooth due to dense networks, and RDS turned out to be efficient in identifying and surveying the Polish population in Reykjavik and its surroundings.

English language publications

Wojtyńska, A. 2011. "From Shortage of Labour to Shortage of Work: Polish Unemployed Immigrants in Iceland", in Ása Guðný Ásgeirsdóttir, Helga Björnsdóttir, and Helga Ólafs (eds), Þjóðarspegillinn. Rannsóknir í félagsvísindum XII. University of Iceland.

Wojtyńska, A. 2011. "Polish Workers in the Capital Area of Iceland", in Sveinn Eggertsson, and Ása G. Ásgeirsdóttir (eds), Þjóðarspegillinn. Rannsóknir í félagsvísindum XIII. University of Iceland.

Polonia in Dublin

The Polonia in Dublin survey is a research project developed by the Department of Sociology/Trinity Immigration Initiative, Trinity College, Dublin. It is a survey of Polish immigrants in the Greater Dublin Area carried out in 2009/2010. The project aimed to explore the use of RDS for Polish migrants in that area, and to study their working conditions, occupational mobility, networks and leisure activities. A total of 623 migrants were interviewed, and overall, RDS proved to be an efficient method for this population. The study was financed by the Irish Research Council for the Humanities and Social Sciences (IRCHSS).

Publications

Mühlau, P. 2013. "Employment and Earnings Mobility of Polish Migrants in Ireland in the Recession", *Annales Universitatis Paedagogicae Cracoviensis (Studia Sociologica IV)*, Special Issue: Migration, Identity, Ethnicity, 118: 81–94.

Mühlau, P. 2012. "Occupational and Earnings Mobility of Polish Migrants in the Recession", IIIS Discussion Paper No. 413.

Reports (available at http://www.tcd.ie/ERC/projectpolonia.php):

Mühlau, P., M. Kaliszewska, & A. Röder. 2011. *Polonia in Dublin: Preliminary Report of Survey Findings – Demographic overview*. Dublin: Employment Research Centre.

Mühlau, P., M. Kaliszewska, & A. Röder. 2011. *Polonia in Dublin: Polish Migrants' Perceptions of Quality of Life, Earnings and Work*. Dublin: Employment Research Centre.

Sub-Saharan Africans in Morocco

The survey on sub-Saharan francophone and anglophone African migrants in an irregular administrative situation was conducted in Rabat, Morocco in March and April of 2013. Migrants were described as being male or female, aged 18 or older, originating from sub-Saharan countries, living and/or working in Rabat, and residing for three months or more in Morocco. The objective of this survey was to collect baseline data on access and use of healthcare and other services, living situations, sexual and drug use behaviors, access to condoms, history and knowledge of sexually transmitted infections, knowledge about HIV transmission and prevention, HIV counseling and testing, stigma and discrimination as well as prevalence of HIV, syphilis, tuberculosis (TB), and malaria. In addition, this survey used RDS data to estimate the size of the population of sub-Saharan francophone and anglophone migrants. The final samples consisted of 410 francophone migrants and 277 Anglophone migrants. Currently, this data is being used by NGOs and by the Moroccan government to make policy changes for improving healthcare access for these migrants. This survey was carried out by UNAIDS, the Global Fund to Fight AIDS, Tuberculosis and Malaria (GFATM), the National AIDS Control Program, and the National Institute of Hygiene. Funding for the survey was provided by GFATM and technical support was provided by UNAIDS.

Publications

Johnston, L. G. 2013a. *HIV integrated Behavioral and Biological Surveillance Surveys-Morocco 2013: Sub-Saharan Migrants in An Irregular Administrative Situation in Morocco.* UNAIDS, Rabat Morocco. Available at: www.lisagjohnston.com.

THEMIS

THEMIS was a large-scale collaborative project running from 2010 to 2014, coordinated by the International Migration Institute and the University of Oxford, and carried out in collaboration with the Peace Research Institute Oslo, the Erasmus University, Rotterdam, and Centro de Estudos Geográficos, University of Lisbon. The project aim was to study under which conditions initial patterns of migration to a certain destination develop into migration systems, and under which conditions they do not. A particular point of interest was the various networks through which migration systems can develop, whether informally (e.g., family, friends, and acquaintances) or formally (e.g., au pair agencies and financial companies). Based on initial scoping studies, we decided to study migration from Brazil, Morocco, and Ukraine to our four countries of settlement – Norway, Portugal, the Netherlands, and the United Kingdom. We carried out large-scale qualitative and quantitative data collection in both the countries of origin and settlement.

RDS was our chosen sampling method for the questionnaire-based quantitative data collection in the countries of settlement for two main reasons. First, we wished to have a certain degree of uniformity across the different contexts, which, in our case, other sampling methods did not provide. Second, our interest in the links between migrant networks and further migration flows meant that the respondents' estimates of their networks, which, as reiterated throughout this book, RDS is dependent on, would in themselves, provide data for the project. Here we mainly draw on experiences from the London and Oslo RDS studies, where THEMIS contributors to this book (Agnieszka Kubal, University of Oxford; Cindy Horst and Rojan Ezzati, Peace Research Institute, Oslo), were involved in the data collection. For more information visit: http://www.imi.ox.ac.uk/research-projects/themis.

SCIP project studies

The SCIP project studies integration trajectories of new immigrants in four European countries: Germany, the Netherlands, Ireland and Great Britain. Its substantive focus is migrants' socio-cultural integration. In the SCIP project, two cross-national waves of survey data are collected among groups of new immigrants that vary along a number of dimensions, including religion (Catholics and Muslims), social status (medium to high-skill and low-skill migrants) and political identity (EU citizens and non-EU-citizens). In all four countries, recently arrived Poles will be sampled, along with new immigrants from Turkey in Germany; from Morocco, Bulgaria, Surinam, and the Dutch Antilles in the Netherlands; and from Pakistan in the UK. RDS was used to sample Poles and Pakistanis in London.

Publications

Renee, L., J. Salamonska, & L. Platt. 2013. "Accounting for Diversity in Polish Migration in Europe: Motivation and Early Integration". Paper presented at *Examining Migration Dynamics: Networks and Beyond*, Oxford, September 2013.

Platt, L., A. Cleary, T. Frere-Smith, & R. Luthra. 2013. "Sampling Recently Arrived Immigrants in the UK: Exploring the Effectiveness of Respondent Driven Sampling". Paper presented at *European Survey Research Association Annual Meeting*, Ljubljana, July 2013.

Appendix II: Overview of the Development of the Different RDS Estimators, Their Specific Features and the Software Available for Their Use[1]

RDS estimator and type	Required data	Limitations	Variance estimation	Analysis feature	Software available
RDSI Heckathorn (1997, 2002).	Recruitment matrix; self-reported network sizes	Limited to categorical data and by RDS assumptions	Bootstrap	Controls for differences in network sizes, homophily across groups; data smoothing[2] for narrower confidence intervals	Yes: RDSAT, STATA RDS estimator (Schonlau and Liebau, 2010)
RDS I Salganik and Heckathorn (2004).	Recruitment matrix; self-reported network sizes	Limited to categorical variables and by RDS assumptions	Bootstrap	Proof that estimate is asymptotically unbiased; estimate of average group network size	Yes: RDSAT, RDS-Analyst
RDSI Heckathorn (2007). Dual-component estimator	Recruitment matrix; self-reported network sizes	Limited by RDS assumptions	Bootstrap	Allows analysis of continuous variables; controls for differential recruitment	Yes: RDSAT, RDS-Analyst
RDS II Volz and Heckathorn (2008). Probability-based Estimator	Recruitment matrix; self-reported network sizes	Limited to nominal variables and by RDS assumptions	Analytic	Allows analysis of continuous variables; shows convergence between reciprocity and probability-based RDS estimators; data smoothing to control for differential recruitment	Yes: RDS-Analyst

RDS estimator and type	Required data	Limitations	Variance estimation	Analysis feature	Software available
Gile Successive Sampling (2011). Probability-based Estimator	Working estimate of population size, recruitment matrix; self-reported network sizes	Currently limited to categorical variables and by RDS assumptions with exception of the with replacement assumption	Bootstrap	Corrects for finite population effects	Yes: RDS-Analyst
Gile and Handcock Model assisted estimator (2012). Probability-based Estimator	Working estimate of population size, recruitment matrix; self-reported network sizes, self-reported composition of contacts helpful but not necessary	Currently limited to binary variables and by RDS assumptions, with exception of the with replacement and seed dependence assumptions	Bootstrap	Corrects for finite population effects and for some forms of seed bias	To be made available in RDS-Analyst

Notes

1. Parts of this table was adapted from Wenjert C, Heckathorn DD. Respondent-Driven Sampling: Operational Procedures, Evolution of Estimators, and Topics for Future Research. The SAGE Handbook of Innovation in Social Research Methods. 2011. (eds) M. Williams, & P. W. Vogt. Sage Publications, London, UK.

2. Data smoothing is a process whereby the number of cross-group recruitments is averaged so that the recruitment matrix of who recruited whom is symmetric (for more information, see Heckathorn, 2002).

References

Arango, J. 2000. "Explaining Migration: A Critical View", *International Social Science Journal*, 52(165): 283–296.
Bakewell, O., A. Kubal, L. Fonseca, G. Engbersen, A. Esteves, S. Pereira, M. Van Meeteren, E. Snel, C. Horst, J. Carling, & R. Ezzati. 2012. "Using Respondent Driven Sampling For Comparative Migration Research: Lessons From The Themis Project", in *Mafe Conference*.
Bernhardt, A., R. Milkman, N. Theodore, D. Heckathorn, M. Auer, J. DeFilippis, A. Luz González, V. Narro, J. Perelshteyn, D. Polson, & M. Spiller. 2009. *Broken Laws, Unprotected Workers*. New York: National Employment Law Project.
Bernard, H. R., E. C. Johnsen, P. D. Killworth, & S. Robinson. 1991. "Estimating the Size of an Average Personal Network and of an Event Subpopulation: Some Empirical Results", *Social Science Research*, 20(2):109–121.
Bøås, M., & I. Bjørkhaug. 2010. "DDRed in Liberia: Youth Remarginalisation or Reintegration?", *MICROCON Research Working Paper 28*, Brighton: MICROCON.
Bonnie H. Erickson. 1978. "Some Problems of Inference from Chain Data", in Karl F. Schuessler (ed.), *Sociological Methodology*. San Francisco: Jossey-Bass. 276–302.
Brunovskis, A., & L. Bjerkan. 2008. *Research with Irregular Migrants in Norway. "Methodological and Ethical Challenges and Emerging Research Agendas"*. Oslo: Udi.

Castles, S., and M. J. Miller. 2009. *The Age of Migration : International Population Movements in the Modern World*. New York: Guilford Press.

Cherti, M. 2008. *Paradoxes of Social Capital. A Multi-Generational Study of Moroccans in The UK*. Amsterdam: Amsterdam University Press, Imiscoe Dissertations.

Drever, A. I., & O. Hoffmeister. 2008. "Immigrants and Social Networks in a Job-Scarce Environment: The Case of Germany", *International Migration Review*, 42: 425–448.

Eckstein, S. 2009. *The Immigrant Divide: How Cuban Americans Changed The US and Their Homeland*. New York: Routledge.

Efron, B., & R. Tibshirani. 1986. "Bootstrap Methods for Standard Errors, Confidence Intervals, and Other Measures of Statistical Accuracy", *Statistical Science*, 1(1): 54–75.

Epstein, G. 2008. "Herd and Network Effects in Migration Decision-Making", *Journal of Ethnic and Migration Studies*, 34: 567–583.

Evans, A. R., G. J. Hart, R. Mole, C. H. Mercer, V. Parutis, C. J. Gerry, J. Imrie, & F. M. Burns. 2011. "Central and East European Migrant Men Who Have Sex With Men in London: A Comparison of Recruitment Methods", *BMC Medical Research Methodology*, 11: 69.

Feskens, R., J. Hox, G. Lensvelt-Mulders, & H. Schmeets. 2006. "Collecting Data among Ethnic Minorities in an International Perspective", *Field Methods*, 18(3): 284–304.

Friberg, J .H. 2010. "Working Conditions for Polish Construction Workers and Domestic Cleaners in Oslo: Segmentation, Inclusion and the Role of Policy", in Richard Black, Godfried Engbersen, Marek Okólski, & Cristina Pantîru (eds), *A Continent Moving West? Eu Enlargement and Labour Migration from Central and Eastern Europe*. Amsterdam: Amsterdam University Press.

Friberg, J. H. 2012. "Culture at Work: 'Polish Migrants in the Ethnic Division of Labour on Norwegian Construction Sites'", *Ethnic and Racial Studies*, 35(11).

Friberg, J. H. 2012. "The Guest-Worker Syndrome Revisited? Migration and Employment Among Polish Workers in Norway's Capital", *Nordic Journal of Migration Research*, 2(4).

Friberg, J. H. 2012. "The Stages of Migration. from Going Abroad to Settling Down: Post-Accession Polish Migrant Workers in Norway", *Journal of Ethnic and Migration Studies*, 38(10): 1589–1605.

Friberg, J. H. 2013. "The Polish Worker in Norway. Emerging Patterns of Migration, Employment and Incorporation after EU's eastern Enlargement". PhD dissertation. Fafo-report 2013: 06.

Friberg, J., J. Arnholtz, L. Eldring, A. Hansen, & K. Thorarins. 2013/2014. "Nordic Labour Market Institutions and New Migrant Workers: Polish Migrants in Oslo, Copenhagen and Reykjavik", Forthcoming in *European Journal of Industrial Relations*.

Friberg, J. H., og L. Eldring. 2011. "Polonia i Oslo 2010. Mobilitet, arbeid og levekår blant polakker i hovedstaden", *Fafo-rapport*, 2011: 27.

Friberg, J. H., & L. Eldring (eds). 2013. "Labour Migrants from Central and Eastern Europe in the Nordic Countries – Patterns of Migration, Working Conditions and Recruitment Practices", *TemaNord*, 2013: 570.

Friberg, J. H., K. Tronstad, & J. E. Dølvik, 2012. "Central and Eastern European Labour Migration to Norway. Trends, Conditions and Challenges", in *Free Movement of Workers and Labour Market Adjustment, Recent Experiences from Oecd Countries and the European Union*. OECD.

Friberg, J., & G. Tyldum, 2007. *Polonia i Oslo. En studie av arbeids- og levekår blant polakker i hovedstadsområdet*. Fafo-report. Oslo: Fafo. 2007: (nr 27).

Gele, A., E. Johansen, & J. Sundby, 2012. "When Female Circumcision Comes to the West: Attitudes toward the Practice Among Somali Immigrants in Oslo", *BMC Public Health*, 12: 697.

Gile, K. J. 2011. "Improved Inference For Respondent-Driven Sampling Data With Application To HIV Prevalence Estimation", *Journal of the American Statistical Association*, 106: 135–146.

Gile, K. J., & M. S. Handcock. 2010. "Respondent-Driven Sampling: An Assessment of Current Methodology", *Sociological Methodology*, 40: 285–327.

Gile, K. J., L. G. Johnston, M. J. Salganik. 2014. "Diagnostics for Respondent-driven Sampling", *Journal of the Royal Statistical Society*, article first published online: 1 MAY 2014. http://onlinelibrary.wiley.com/doi/10.1111/rssa.12059/pdf

Goel, S., & M. Salganik. 2009. "Respondent Driven Sampling as Markov Chain Monte Carlo", *Statistics in Medicine*, 28(17): 2202–2229.

Grant, R. W. 2006. "Ethics and Incentives: A Political Approach", *American Political Science Review*, 100: 29–39.

Groenewold, G., & R. E. Bilsborrow. 2008. "Design of Samples for International Migration Surveys: Methodological Considerations and

Lessons Learned from a Multi-country Study in Africa and Europe", in C. Bonifazi, M. Okólski, J. Schoorl, & P. Simon (eds), *International Migration in Europe; New Trends and New Methods of Analysis*. Amsterdam: Amsterdam University Press. 293–312.

Groves, R. M., S. Presser, & S. Dipko. 2004. "The Role of Topic Interest in Survey Participation Decisions", *Public Opinion Quarterly*, 68: 2–31.

Groves, R. M., E. Singer, & A. Corning. 2000. "Leverage-Saliency Theory of Survey Participation – Description and an Illustration", *Public Opinion Quarterly*, 64: 299–308.

Hammond, L. 2011. "The Absent but Active Constituency: The Role of the Somaliland UK Community in Election Politics", in P. Mandaville, & T. S. Lyons (eds), *Politics From Afar: Transnational Diasporas and Networks*. London: C. Hurst & Co.

Handlin, O. 1952. *The Uprooted*. Boston: Little Brown.

Hansen, J. A., & N. W. Hansen. 2009. "Polonia i København. 'Et studie af polske arbejdsmigranters', løn-, arbejds- og levevilkår i Storkøbenhavn", *Lo-dokumentation*, 1.

Heberlein, T. A., & R. Baumgartner. 1978. "Factors Affecting Response Rates to Mailed Questionnaires: A Quantitative Analysis of the Published Literature", *American Sociological Review*, 43: 447–467.

Heckathorn, D. 1997. "Respondent-Driven Sampling: A New Approach to the Study of Hidden Populations", *Sociological Problems*, 44(Suppl 2): 174–199.

Heckathorn, D. 2002. "Respondent Driven Sampling II: Deriving Valid Population Estimates from Chain-Referral Samples of Hidden Populations", *Sociological Problems*, 49(Suppl 1): 11–34.

Heckathorn, D. 2007. "Extensions of Respondent Driven Sampling: Analyzing Continuous Variables and Controlling For Differential Recruitment", *Sociological Methodology*, 37(1): 151–207.

International Organization for Migration (IOM). 2001. "International Migration, Racism, Discrimination and Xenophobia", Accessed on November 30, 2013: http://www.unesco.org/most/migration/imrdx.pdf.

Johnston, L. G. 2008. *Behavioral Surveillance: Introduction to Respondent Driven Sampling* (Participant Manual). Centers for Disease Control and Prevention, Atlanta, GA. http://globalhealthsciences.ucsf.edu/PPHG/surveillance/other_modules.html.

Johnston, L. G. 2011. "The Role of Formative Research", in *Design, Implementation and Analysis: An Exploration of Respondent Driven Sampling*. London, June 16, 2011. www.lisagjohnston.com

Johnston, L. G. 2013a. *HIV Integrated Behavioral and Biological Surveillance Surveys-Morocco 2013: Sub-Saharan Migrants in An Irregular Administrative Situation in Morocco*. UNAIDS, Rabat, Morocco. www.lisagjohnston.com.

Johnston, L. G. 2013b. *Introduction to Respondent Driven Sampling*. Geneva: WHO. http://applications.emro.who.int/dsaf/EMRPUB_2013_EN_1539.pdf

Johnston, L. G., R. Khanam, M. Reza, S. Khan, S. Banu, M. Alam, M. Rahman, & T. Azim. 2007. "The Effectiveness of Respondent Driven Sampling for Recruiting Males Who Have Sex with Males in Dhaka, Bangladesh: A Pilot Study", *AIDS and Behavior*, 2(2): 294–304.

Johnston, L. G., & M. Malekinejad. 2014. "Chapter VII: Respondent-Driven Sampling for Migrant Populations", in M. Schenker, X. Castaneda, & A. Rodriguez Lainz (eds), *Synopsis of Migration and Health: A Research Methods Handbook*. University of California Press, California: Berkeley.

Johnston, L. G., M. Malekinejad, M. Rifkin, G. W. Rutherford, & C. Kendall. 2008. "Implementation Challenges to Using Respondent-Driven Sampling Methodology for HIV Biological and Behavioral Surveillance: Field Experiences in International Settings", *AIDS and Behavior*, 12(Suppl 1): 131–141.

Johnston, L. G., S. Whitehead, M. Simic, & C. Kendall. 2010. "Formative Research To Optimize Respondent Driven Sampling Surveys Among Hard To Reach Populations in HIV Behavioral and Biological Surveillance: Lessons Learned From Four Case Studies", *AIDS Care*, 22(6): 784–792.

Kalton, G. 2001. "Practical Methods for Sampling Rare and Elusive Populations", in Proc. A. Meet. American Statistical Association. Alexandria: American Statistical Association.

Kalton, G., & D. W. Anderson. 1986. "Sampling Rare Populations", *Journal of the Royal Statistical Society. Series A*, 149(1): 65–82.

Karon, J. M., & C. Wejnert. 2012. "Statistical Methods for the Analysis of Time-Location Sampling Data", *J Urban Health*, 89(3): 565–586.

Kasinitz, P. 2008. *Inheriting the City: The Children of Immigrants Come of Age*. Cambridge, MA: Harvard University Press.

Khamsiriwatchara, A., P. Wangroongsarb, J. Thwing, J. Eliades, W. Satimai, C. Delacollette, & J. Kaewkungwal. 2011. "Respondent-driven Sampling on the Thailand-Cambodia Border. I. Can Malaria Cases be Contained in Mobile Migrant Workers?", *Malaria Journal*, 10: 120.

King, R. 2002. "Towards a New Map of European Migration", *Population Space and Place*, 8(2): 89–106.
Kubal, A., & R. Dekker. 2011. "Consequences of Immigrant Inter-Wave Dynamics for Migration Processes: Ukrainians in the Netherlands and the United Kingdom", IMISCOE Conference, Warsaw, September 2011.
Leskovec, J., & E. Horvitz. 2008. "Planetary-Scale Views on an Instant-Messaging Network", Proceedings of the 17th international conference on World Wide Web. ACM.
Luthra, R., A. Cleary, & T. Frere-Smith. 2013. "Sampling Recently Arrived Immigrants in the UK: Exploring the Effectiveness of Respondent Driven Sampling", Paper presented at the 5th Esra Conference, Ljubljana, Slovenia.
Macklin, R. 1981. " 'Due' and 'Undue' Inducements: On Paying Money to Research Subjects", *IRB Ethics and Human Research*, 3(5): 1–6.
Malekinejad, M., L. G. Johnston, C. Kendall, L. Kerr, M. R. Rifkin, & G. W. Rutherford. 2008. "Using Respondent-Driven Sampling Methodology for HIV Biological and Behavioral Surveillance in International Settings: A Systematic Review", *AIDS and Behavior*, 12: 105–130.
Massey, D., J. Arango, G. Hugo, A. Kouaouci, A. Pellegrino, & J. Taylor. 1998. *Worlds in Motion – Understanding Migration at the End of the Millennium*. Oxford: Clarendon Press.
McCarty, C., P. D. Killworth, H. R. Bernard, E. Johnsen, & G. A. Shelley. 2001. "Comparing Two Methods for Estimating Network Size", *Human Organization*, 60: 28–39.
McCormick, T. H., M. Salganik, & T. Zheng. 2010. "How Many People Do You Know? Efficiently Estimating Personal Network Size", *Journal of the American Statistical Association*, 105(489): 59–70.
Mckenzie, D. A. S., & J. Marcin. 2007. "Migration, Remittances, Poverty, and Human Capital: Conceptual and Empirical Challenges", *World Bank Policy Research Working Paper*, (online), No. 4272.
McKenzie, D. J., & J. Mistiaen. 2009. "Surveying Migrant Households. A Comparison of Census-Based, Snowball and Intercept Point Surveys", *Journal of the Royal Statistical Society: Series A*, 172(2): 339–360.
McPherson, M., L. Smith-Lovin, & J. M. Cook. 2001. "Birds of a Feather: Homophily in Social Networks", *Annual Review of Sociology*, 27: 415–444.
Milgram, S. 1967. "The Small World Problem", *Psychology Today*, 1(1): 60–67.

Milkman, R., A. L. González, & V. Narro. 2010. *Wage Theft and Workplace Violations in Los Angeles. The Failure of Employment and Labor Laws for Low-Wage Workers.* Institute for Research on Labor and Employment. Los Angeles: University of California.

Montealegre, J. R., L. G. Johnston, C. Murrill, & E. Monterroso. "Respondent Driven Sampling for HIV Biological and Behavioral Surveillance in Latin America and the Caribbean", *AIDS and Behavior*, 2013. (epub ahead of print).

Montealegre, J. R., J. M. Risser, & B. J. Selwyn. 2012. "Prevalence of HIV Risk Behaviors among Undocumented Central American Women in Houston, Texas", *AIDS and Behavior*, 16(6): 1641–1648.

Montealegre, J. R., J. M. Risser, B. J. Selwyn, S. A. McCurdy, & K. Sabin. 2013. "Effectiveness of Respondent Driven Sampling to Recruit Undocumented Central American Women in Houston, Texas for an HIV Behavioral Survey", *AIDS and Behavior*, 17(2): 719–727.

Montealegre, J. R., J. M. Risser, B. J. Selwyn, K. Sabin, & S.A. McCurdy. 2011. "HIV Testing Behaviors among Undocumented Central American Women in Houston, Texas", *Journal of Immigrant and Minority Health*, 14(1): 116–123.

Muhib, F. B., L. S. Lin, A. Stueve, R. L. Miller, W. L. Ford, W. D. Johnson, & P. J. Smith. 2001. "A Venue-Based Method for Sampling Hard-to-Reach Populations", *Public Health Reports*, 116(Suppl 1): 216–222.

Mühlau, P. 2012. "Occupational and Earnings Mobility of Polish Migrants in the Recession", *IIIS Discussion Paper*, No. 413.

Mühlau, P. 2013. "Employment and Earnings Mobility of Polish Migrants in Ireland in the Recession", *Annales Universitatis Paedagogicae Cracoviensis (Studia Sociologica IV)*, Special Issue: "Migration, Identity, Ethnicity", 118: 81–94.

Mühlau, P., M. Kaliszewska, & A. Röder. 2011. *Polonia in Dublin: Preliminary Report of Survey Findings – Demographic Overview.* Dublin: Employment Research Centre.

Palloni, A., D. Massey, M. Ceballos, K. Espinosa, & M. Spittel. 2001. "Social Capital and International Migration: A Test Using information On Family Networks", *The American Journal of Sociology*, 106: 1262–1298.

Piore, M. J. 1979. *Birds of Passage: Migrant Labor and Industrial Societies.* Cambridge: Cambridge University Press.

Portes, A., & R. G. Rumbaut. 2001. *Legacies: The Story of the Immigrant Second Generation.* Berkeley: University of California Press.

Ramirez-Valles, J., D. D. Heckathorn, R. Vázquez, R. M. Diaz, & R. T. Campbell. 2005. "From Networks to Populations: The Development and Application of Respondent-Driven Sampling Among IDUs and Latino Gay Men", *AIDS and Behavior*, 9(4): 387–402.
Risser, J. M., & J. R. Montealegre. 2013. "Comparison of Surveillance Sample Demographics Over Two Cycles of the National Hiv Behavioral Surveillance Project – Houston, Texas", *AIDS and Behavior*, Published online, August 2, 2013.
Rodriguez, L. 2005. "Generations and Motivations: Russian and Other Former Soviet Immigrants in Costa Rica", *International Migration*, 43(4): 147–165.
Rodriguez, L. 2009. "Economic Adaptation and the Self-employment Experience of Nigerian Immigrants in New York City". PhD dissertation. The Pennsylvania State University.
Roose, H., J. Lievens, & H. Waege. 2007. "The Joint Effect of Topic Interest and Follow-Up Procedures on the Response in a Mail Questionnaire – An Empirical Test of the Leverage-Saliency Theory in Audience Research", *Sociological Methods & Research*, 35: 410–428.
Salganik, M. J., & D. D. Heckathorn. 2004. "Sampling and Estimation in Hidden Populations Using Respondent-Driven Sampling", *Sociological Methodology*, 34: 193–240.
Schonlau, M., & E. Liebau. 2010. "RDS-A Stata Program for Respondent-Driven Sampling", Stata Users Group.
Semaan, S., S. Santibanez, R. S. Garfein, D. D. Heckathorn, & D. C. Des Jarlais. 2008. "Ethical and Regulatory Considerations in HIV Prevention Studies Employing Respondent-Driven Sampling", *International Journal of Drug Policy*, 20(1):14–27.
Solomon, S. S., G. M. Lucas, D. D. Celentano, F. Sifakis, & S. H. Mehta. 2013. "Beyond Surveillance: A Role For Respondent-Driven Sampling in Implementation Science", *American Journal of Epidemiology*, 178: 260–267.
Stoop, I. A. L. 2004. "Surveying Nonrespondents", *Field Methods*, 16(1): 23–54.
Townsend, L., M. Giorgio, Y. Zembe, M. Cheyip, & C. Mathews. 2014. "HIV Prevalence and Risk Behaviours among Foreign Migrant Women Residing in Cape Town, South Africa", *AIDS and Behavior*. Under review.
Þórarinsdóttir, H., & A. Wojtyńska. 2011. *Polonia Reykjavik 2010: Preliminary Report*. Reykjavik: MIRRA.

Tyldum, G. 2012. "Ethics or Access? Balancing Informed Consent Against the Application of Institutional, Economic or Emotional Pressures in Recruiting Respondents for Research", *International Journal of Social Research Methodology*, 15: 199–210.

United Nations, Department of Economic and Social Affairs (UNDESA). 2013. *The Number of International Migrants Worldwide Reaches 232 Million*. Population Facts. No. 2013/2.

United Nations, Office of the High Commissioner on Human Rights (UNHCR). 2013. *High-Level Dialogue on International Migration and Development*. New York. Accessed on November 30, 2013 at: http://www.ohchr.org/Documents/Issues/SRMigrants/A-68-283.pdf.

Van, L. I., & V. Bilger. 2012. "Ethical Challenges in Research with Vulnerable Migrants", in C. Vargas-Silva (ed.), *Handbook of Research Methods in Migration*. Cheltenham: Edward Elgar. 451–466.

Volz, E., & D. D. Heckathorn. 2008. "Probability Based Estimation Theory for Respondent Driven Sampling", *Journal of Official Statistics*, 24: 79–97.

Watters, J. K. & P. Biernacki. 1989. "Targeted Sampling: Options for the Study of Hidden Populations", *Sociological Problems*, 36: 416–430.

Wejnert, C. 2009. "An Empirical Test of Respondent-Driven Sampling: Point Estimates, Variance, Degree Measures, and Out-of-Equilibrium Data", *Sociological Methodology*, 39: 73–116.

Wejnert, C., & D. D. Heckathorn. 2008. "Web-Based Network Sampling: Efficiency and Efficacy of Respondent-Driven Sampling for Online Research", *Sociological Methods & Research*, 37(1): 105–134.

White, R. G., A. Lansky, S. Goel, D. Wilson, W. Hladik, A. Hakim, & S. D. Frost. 2012. "Respondent Driven Sampling – Where We Are and Where Should We Be Going?", *Sexually Transmitted Infections*, 88: 397–399.

Wimmer, A., & N. Glick-Schiller. 2002. "Methodological Nationalism and Beyond: Nation-State Building, Migration and The Social Sciences", *Global Networks*, 2: 301–334.

Winship, C., & L. Radbill. 1994. "Sampling Weights and Regression Analysis", *Sociological Methods and Research*, 23(2): 230–257.

Wojtyńska, A. 2011. "From Shortage of Labour to Shortage of Work: Polish Unemployed Immigrants in Iceland", in Ása Guðný Ásgeirsdóttir, Helga Björnsdóttir, & Helga Ólafs (eds), *Þjóðarspegillinn. Rannsóknir í félagsvísindum XII*. University of Iceland.

Zhang, S. X. 2012. *Trafficking of Migrant Laborers in San Diego County: Looking for a Hidden Population*. San Diego, CA: San Diego State University.

Index

altruism, 51
analysis software, 85
anonymous participation, 4, 13
appointment based, 65, 66, 67, 68

bartering of coupons, 52, 55, 97
bias, 38, 43, 92, 96
　reducing bias, 12
boosting participation, 50
bottlenecks, xii, 23, 24, 40, 46, 54
　analysis of, 96

Central American Women in Houston, 22, 23, 32, 63, 67, 68, 70, 74, 77, 79, 81, 93, 101
clustering, xii, 23
coupon logs, 75, 77
coupon numbering errors, 75
coupons, xii, 11, 70, 72
　activation dates, 79
　number of coupons, 66, 79, 80
　reducing number of coupons, 80
　section for redeeming secondary incentive, 58, 72
cross-site recruitment, 75

differential recruitment, 88, 94

ending recruitment, 80
equilibrium, xii, 16, 38, 90, 91

estimators, 89, 96
ethical issues
　anonymity, 81
　confidentiality, 81
　ethical review, 56, 82
　incentives, 59
　risk of harm, 60
　safety, 82
　vulnerable populations, 25, 60, 70, 81
exporting weights, 96

fixed-site approach, 66, 67, 68
follow-up interview, 58
Foreign Migrants in Ukraine, 24, 102
formative assessment, 25, 31, 40, 52, 64, 74

geographic boundary, 31

homophily, xii, 15, 40, 77, 93
hours of operation, 69
household surveys, 10

incentives, 11, 12, 61
　amount, 52, 60
　a positive interview experience, 57
　collection of secondary incentive, 58
　cultural differences, 58
　distribution, 58

Index 125

incentives – *continued*
  non-monetary, 56
  primary incentive, xiii, 51
  RDS without material incentives, 56
  secondary incentive, xiii, 51
  security concerns, 58
  stratified incentives, 54
  too high, 55
  too low, 55, 77

key informant interviews, 25

mandatory information, 28
Markov chain, 15, 99
masquerading, 55, 79
migrant populations
  access, 3
  diversity, 39
  ethical concerns, 25
  hard to reach, 3, 10
  identification, 3, 29
  lack of sampling frame, 4
  migrant networks, 20
  need for data, 2
  networks, 21
Migrants in Warsaw, 24, 102
mixed methods approach, 25
mobile sites, 67, 69, 82
motivation, 3, 51
multivariate analysis, 96

Nigerians in New York City, 57, 104
number of coupons, 14

parallel monitoring, 64, 77, 80
peer pressure, 12, 51
peer-to-peer recruitment, 11
personal network size, xiii, 14, 28, 36, 67
PNS question, 36, 96
  placement, 28
  time period, 31
Polonia in Dublin, 106
Polonia in Oslo, 22, 23, 24, 29, 40, 44, 52, 55, 57, 59, 68, 69, 78, 80, 79, 93, 98, 104

Polonia in Reykjavik, 40, 46, 79, 105
population count, 88
probability sampling, 4

random recruitment within network, 14
random walk, 100
rapid recruitment, 78
RDS Analysis Tool (RDSAT), 87
RDS Analyst, 87
RDS-I estimator, 88, 96
RDS-II estimator, 88, 96
recently arrived migrants, 22, 29, 63
reciprocal ties, 13, 18, 30
recruiters, xii
recruitment, 34, 38, 74
  across bottlenecks, 54
  script for recruitment, 46
  slow recruitment, 77
recruitment chains, xiii, 11, 12, 77
recruitment matrix, xiii
recruits, xii
repeat participation, 79

safety, 82
sampling frame, 10
screening, 53, 69, 80
seasonal migration, 21, 35, 74
seed dependence, 12, 15, 91
seeds, xiii, 11, 24, 64, 77
single network component, 14, 21, 22, 95
slow recruitment, 55, 64, 66
snowball sampling, 10
social categories, 20
social groups, 20, 21
  naturally occuring, 21
staffing, 69, 70
start date, 74
STATA, 87
strong ties, 30
Sub-Saharan Africans in Morocco, 11, 41, 47, 66, 81, *88*, 93, 107
Successive Sampling (SS) estimator, 88, 96

survey sites, 57, 64, 67
    multiple sites, 68

targeted sampling, 10
target population
    definition, 21, 29
    time of arrival, 23
THEMIS, xiv, xv, 20, 23, 25, 40, 42, 43,
    45, 46, 52, 58, 63, 65, 66, 69, 70, 72,
    74, 77, 78, 81, 82, 108
time-location sampling, 10
training of staff, 33, 57, 58, 81
transnational mobility, 21

trust, 3, 43, 77, 81

unadjusted sample statistics, 85
undocumented migrants, 63

variance, 90
volunteerism, 51

walk-in, 67
waves, xiii, 11
weak ties, 30
web-based approach, 67
with replacement sampling, 14

The manufacturer's authorised representative in the EU is Springer Nature Customer Service Centre GmbH, Europaplatz 3, 69115 Heidelberg, Germany. If you have any concerns regarding our products, please contact ProductSafety@springernature.com

Printed and bound by CPI Group (UK) Ltd, Croydon, CR0 4YY
23/03/2026
02076402-0018